The Struggle for the Land

The Struggle for the Land

Indigenous Insight and

Industrial Empire

in the Semiarid World

Edited by

Paul A. Olson

Published by the University of
Nebraska Press, Lincoln and London,
for the Center for Great Plains Studies,
University of Nebraska–Lincoln

Chapter 10, J. Baird Callicott's "American Indian
Land Wisdom," is reprinted with minor revisions
from J. Baird Callicott, *In Defense of the Land Ethic:
Essays in Environmental Philosophy*, and is published
here by permission of the State University of New
York Press. © 1989 State University of New York.

Library of Congress Cataloging-in-Publication Data
The struggle for the land: indigenous insight and
industrial empire in the semiarid world /
edited by Paul A. Olson. p.   cm.
Published for the Center for Great Plains
Studies, University of Nebraska–Lincoln.
ISBN 0-8032-3555-0
1. Indigenous peoples. 2. Land use.
3. Land tenure. 4. Arid regions agriculture.
5. Arid regions ecology. 6. Industrialization.
I. Olson, Paul A. II. University of Nebraska–
Lincoln. Center for Great Plains Studies.
GN380.S86   1990   306.3'2—DC20
89-22422   CIP

In Honor of Reuben Snake,

chairperson of the Winnebago Tribe

Syd Beane of the Santee Sioux Tribe,

director of the Lincoln Indian Center

Leaders in the struggle for their people's

lands, traditions, and economic rights

Contents

Preface

✦

The essays in this book largely originated in the 1986 symposium of the Center for Great Plains Studies of the University of Nebraska–Lincoln. In 1984 several colleagues—David Wishart, Erwin Goldenstein, Frederick Luebke, Peter Bleed, Elizabeth Grobsmith, Elaine Jahner, Gary Moulton, Frances Kaye, Webster Robbins, Syd Beane, Jon Nelson, Francis Blythe, and Wallace Coffey—and I were asked to organize the tenth annual Center for Great Plains Studies symposium on the theme "The Meaning of the Plains Indian Past for Present Culture." One of the subthemes we determined to pursue was the meaning of past indigenous adaptive strategies and land use for present management of land and resources in semiarid regions, many of which are in trouble. The Center for Great Plains Studies, which organizes these annual symposia, is an interdisciplinary agency of the University of Nebraska designed to coordinate research, teaching, and service activities contributing to an understanding of the Great Plains and similar regions across the globe, and so the subtheme seemed appropriate to its mission.

For the culture-and-environment section of the symposium that resulted, the committee endeavored to secure analysts who knew what had happened, after European-based occupation, to several groups of indigenous peoples and to the surrounding semiarid worlds in which they had moved, and we found persons who could treat of such matters in the Great Plains, Alaska, Africa, Australia, and Central Asia. We did not secure a scholar who had looked at the Araucanian regions in South America, for reasons to be explained later, and we omitted the Sahel because it had been so intensively studied in the mid-1980s and most of the findings were similar to those presented in this book. After the conference, the World Commission on Environment and Development published *Our Common Future*, which endorsed much that major symposium scholars had also found, but primarily with an eye to other than semiarid regions. A book that examined what had happened to semiarid regions and peoples since the European-based occupation could, we thought, add to the insights to be found in *Our Common Future*.

To make this volume, we put together the best of the essays presented on the theme in the 1986 symposium and commissioned two others. All were revised by their authors in light of the work's total thrust, rigorously critiqued by the University of Nebraska Press's reviewers, and again revised. At the same time, we received help with the revisions from a number of people—Louise Fortmann of the Anthropology Department of the University of California at Berkeley, Robert Hitchcock of the University of Nebraska Anthropology Department, David Lewis of the University of Nebraska–Lincoln Agronomy Department, Marty Strange of the Center for Rural Affairs, Clyde Milner of the University of Utah, and Fred Luebke of the Center for Great Plains Studies. I have also received useful editorial counsel from Rosalind Carr, Roma Rector, and Marylee Yetter, who helped produce the manuscript. In the final redrafting, Carol Berigan was of invaluable assistance. Much of the work's conceptual structure came out of conversations with my daughter, Ingrid Olson, a student of applied anthropology, and my wife, Elizabeth Olson, a student of international affairs.

Books like this are not put together without funds. The primary support for its creation and for the conference that led to it came from the John Ben Snow Memorial Trust. Other groups who contributed to the creation of the conference and the book include the University of Nebraska's Center for Great Plains Studies Research Improvement Fund, the National Endowment for the Humanities, the InterNorth Foundation, the Canadian Consulate-General, the Canadian Department of External Affairs, the University of Nebraska International Programs division, the Dean of the College of Arts and Sciences, and the University of Nebraska–Lincoln Chancellor's Office. To all these entities, I am grateful.

Any errors in this book are my responsibility. May Mercury, god of scholars, thieves, and merchants, look on them with a tolerant eye.

Introduction

Paul A. Olson

At the end of the 1887 council where his people were told to learn farming in the semidesert region east of the Wind River Mountains, the Shoshone chief Washakie exploded with, "God damn a potato!" (Arrowsmith 1971, 3).[1] To Arrowsmith, Washakie's "winged words" were an indication that buffalo hunters understood they had better things to do than raise potatoes. They had dreams to dream, which bore their identity, and a membership in a great natural community to sustain.

Arrowsmith may have been right. But by 1887 Washakie had been on his reservation for some years, and he may have meant something quite prophetic instead—that the potato would never replace the buffalo in dryland areas, that agriculture as Europeans practiced it could not thrive in the part of Wyoming he was being told to farm. In the same passage Arrowsmith quotes Smohalla, the Nez Perce leader, as saying, "You ask me to cut grass and make hay and sell it and be rich like the white man. But how dare I cut my Mother's hair? Shall I take a knife and tear my Mother's breast? Then when I die, she will not take me to her bosom to rest" (Arrowsmith 1971, 3–4).

The question is, Did Smohalla and Washakie say these things for religious or practical reasons? Clearly, many indigenous peoples on this continent and others resisted cutting open the earth and making their living in the new way dictated by the "Wasichus" or "fat eaters" from Europe. It is not clear that they did so only because of their dreams, their identities, or their mystical sense of natural community. They may have done so partly because they had a pragmatic sense that semiarid lands should be used as they had used them and that other ways would prove impractical over time. They may have felt that their spiritual connections to the land and practical uses of it were related.

Often these statements of resistance to European ways are, in Western accounts, treated as heroic statements by the noble defeated rather than as pragmatic advice from seasoned survivors. That they are something other than noble rhetoric is suggested by the environmental degradation of semiarid regions subjected to a century of European-based development. In any case, the relationship between the buffalo hunter and the potato eater, between their respective dreams, identities, and senses of natural communities—the relationship between indigenous peoples and industrial empire—is the theme of this book.

The essays concern the effect of European-based peoples in the past century and a half on peoples living in semiarid lands on many continents, these groups' reciprocal effect on European-based peoples, and the effect of both on Smohalla's "mother"—the land.[2] They look at industrial/indigenous relationships in the lands of the North American Plains Indians, the Australian Aborigines, the Kazakhs in the USSR, the Maasai in Kenya, several groups in southern Africa, and Alaskan and Lapp (Saami) native peoples. The primary "empires" affecting these people were England and the United States of America, sharing common expansionist traditions, and Russia or the Soviet Union, with its quite different traditions. We omit Latin American peoples coming from like regions because genocide has destroyed their larger groups.

Recently the World Bank and other development agencies have given special attention to the deleterious environmental effects of international industrial development because so many projects have created deserts rather than gardens. Though the authors in this volume argue that pastoral nomadic and hunter-gatherer peoples who live in semiarid regions are rational—that in the main they have used their worlds in sustainable ways that have permitted them to produce the goods necessary to their long-term survival—their view is not the dominant one. Another view having great currency in the development literature, which Garrett Hardin expresses in writing about the "tragedy of the commons," asserts that where people have the opportunity to make private gain from "common" resources, they will over-use them for short-term advantage until they destroy all available resources, unless resources are so plentiful and population is so sparse that destruction is impossible (Hardin and Baden 1977, 16–30). Hardin's argument derives from William Lloyd's nineteenth-century analysis of the last days of the degradation of the commons in England, when the legal structure and role system that once sustained the practical use of common pasture and woodland had virtually disappeared and many commons *were* degraded. But Hardin uses Lloyd's concept to urge that, in the contemporary world, common resources—such as land, air, water, the wealth of the sea, and the right to reproduce—must either be made private property so that the profit motive will ensure responsibility or be subject to rigorous national or international control. Following Hardin, policymakers active in developing areas where environmental problems are likely have often stressed privatization or strict external regulation of access to resources.

However, recently McCay and Acheson and other authors of *The Question of the Commons* have demonstrated, contrary to Hardin, that both preindustrial Western societies and modern tribal ones control access to common resources through the tools of law, custom, myth, and role system, and that what Hardin sees as the "tragedy of the commons" exists only where such constraints have been destroyed or where new constraint systems are needed because new industrial technologies reach into previously untouched areas (McCay and Acheson 1987). McCay and Acheson's argument applies mostly to relatively remote societies, geographically or psychologically considered. In transitional societies, once the old constraint systems have been broken, efforts to establish a new "commons" through cooperatives, tribal farms, group ranches, and the like have often not fared well. But efforts to follow the privatization solution have not succeeded either in drier regions, because private-property land systems seem to preclude flexibility in periods of drought or extraordinary plenty. Thus two recurrent problems present themselves in the contradictory solutions—the destruction of flexibility in the creation of private property and the elimination of human responsibility through the creation of commons. Of course, human communities throughout the world are working on various sorts of management compromises that call for both corporate action and individual responsibility, compromises to be treated later in the book.

In considering what has been and should be done in dryland regions, this book's discussion turns on three key terms: *semiarid, indigenous peoples,* and *industrialization*. By *semiarid* we mean lands receiving on the average 10–20 inches (250–500 millimeters) of rainfall annually (the season of rainfall and evapotranspiration rates may also affect what is considered semiarid). In general, semiarid climax that has not been degraded remains grassland in temperate areas, semiarid savanna in tropical ones, and taiga or tundra in arctic regions (see map 1.1). But great variation exists. The length of the dry season may be crucial. Since the territories treated here have continental climates at the mercy of prevailing wind currents, they are subject to great variability and may periodically become subhumid or arid. Within a specific temperate grassland region, for example, bottomlands may be covered with tallgrasses characteristic of subhumid regions while hilltops are nearly desert (e.g., Harris 1980, 3–31). All the areas discussed required of their preindustrial dwellers foraging, pastoralism, or agropastoralism.

By *indigenous peoples* we mean tribal peoples—peoples who had lived in a region a century or more before the advent of European-based industrial empire, often using a nomadic or seminomadic style of living to obtain resources from the land. The subjects of the investigations are what we consider representative groups. Other groups might be substituted for the Plains Indians of the United States and Canada, the Inuit and Saami of Alaska and Europe, the Kazakhs of Soviet Central Asia, the Aborigines of central Australia, the Maasai of Kenya, and a group of tribes, especially the Tswana, living in the southern portion of Africa (map 1.2). Though slightly different stories might then emerge, we think the stories given are typical enough to permit useful comparisons and tentative general hypotheses. The authors had already done some comparative indigenous studies, and for this book we asked them to think further about the relation between industrial and indigenous adaptations in semiarid regions: in particular, the issues of changes in land use, institutional development before and after colonization, and changes in "dream" (or religion and ethical code). These issues are the topics of the last three sections of the book. Though we find trends in the history of these relationships, we do not discover universal patterns, and we recognize the importance of Spicer's caveat that it is "meaningless to posit . . . invariant sequences" in the history of interaction between indigenous peoples and Europeans "without explicit statement of the conditions of

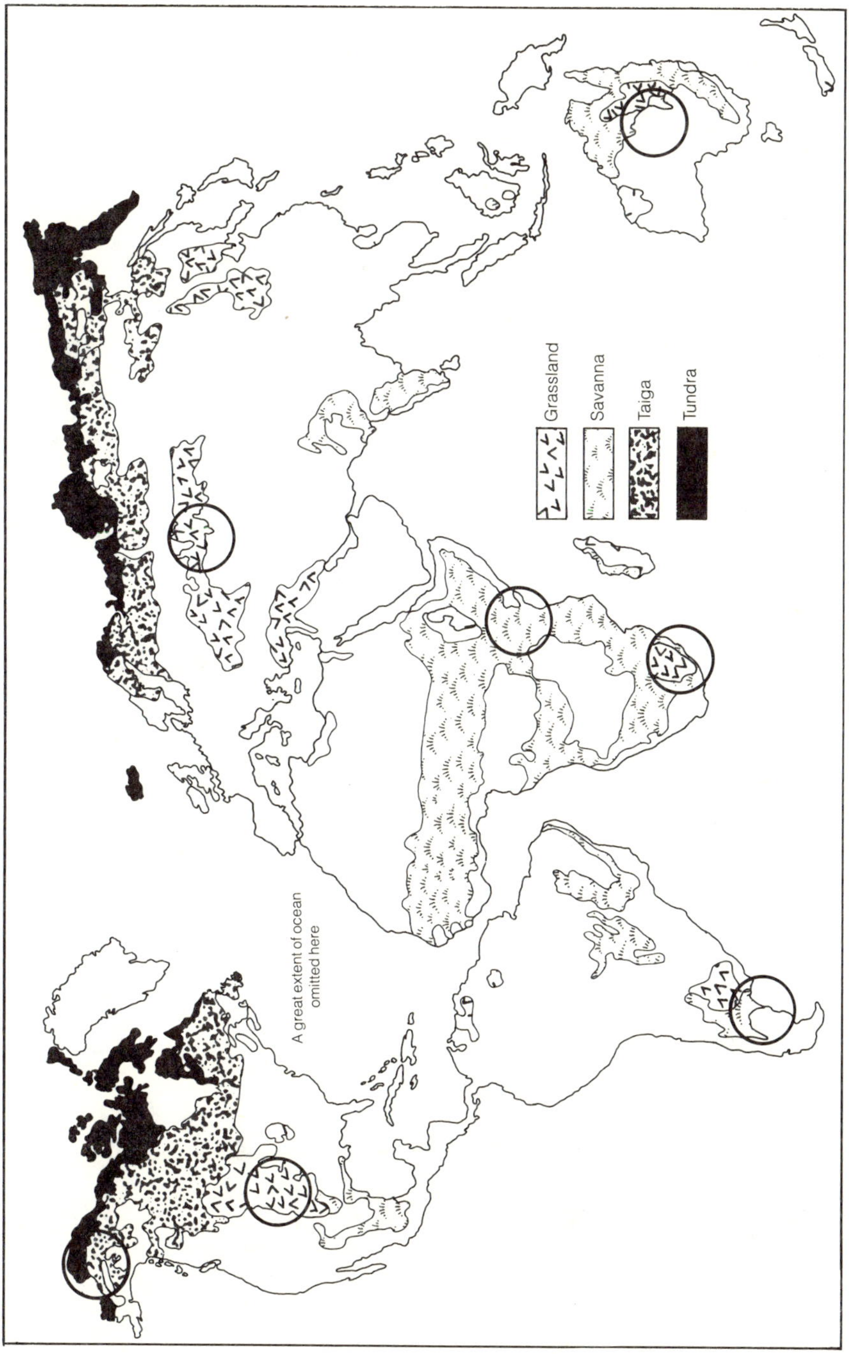
Grassland
Savanna
Taiga
Tundra
A great extent of ocean omitted here

Map 1.1 Semiarid-land vegetation in approximate areas occupied by peoples discussed in this book.

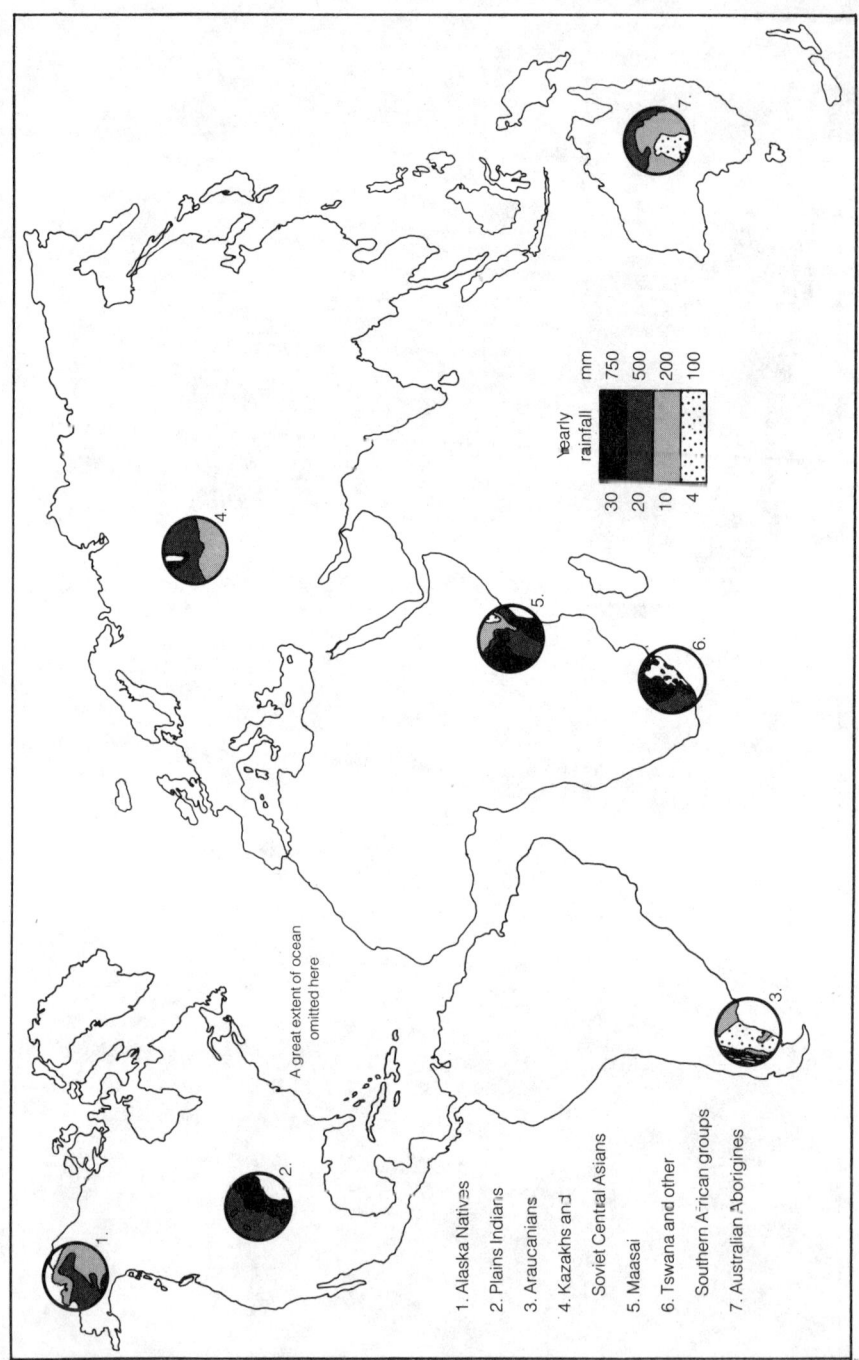

Map 1.2 Semiarid regions and peoples discussed in this book.

Legend (yearly rainfall):

mm
750 — 30
500 — 20
200 — 10
100 — 4

1. Alaska Natives
2. Plains Indians
3. Araucanians
4. Kazakhs and
   Soviet Central Asians
5. Maasai
6. Tswana and other
   Southern African groups
7. Australian Aborigines

A great extent of ocean omitted here

6

contact as well as of the nature of the cultures in contact" (Spicer 1961, 54).

Since this volume contains work primarily by humanities scholars, value judgments and perspectives gathered from within the several semiarid civilizations are to be expected (Pike 1954; Pike and Brend 1976). The groups studied are not equally "indigenous." Some of them, for example the Australian Aborigines, had lived in their land for thousands of years before white contact, whereas some of the Plains Indians had lived on the Plains only a hundred years or so before similar contact. But either time is long enough to adapt. That is, each group studied had formed a set of normative habits that permitted it to win a living from the ecosystem that surrounded it and to transmit those habits to future generations—through actions, institutions, and narratives or ceremonies serving a religious-educational purpose. Each had found a niche or a set of niches that permitted it to survive.[3]

Finally, by *industrialization* we mean both the creation of new technologies and the development of the institutions of industrial capitalism (though the latter was aborted in the Soviet Union). The new technologies included the introduction of the common nonhuman, nonanimal energy sources associated with the industrial revolution—water and fossil-fuel power, especially the steam engine and the mechanized factory and mine—but also long-range weaponry like the breech-loading rifle.[4] Most important, they mean the contrivances of industrial agriculture—the row crop drills, mechanical cultivators, elaborate drainage and irrigation schemes, and scientific stock breeding developed in England in the seventeenth and eighteenth centuries (Kerridge 1967); Mecke's early nineteenth-century threshing machine; the 1840s chemical analyses and addition of nitrogen and phosphate to soil culture; and the 1840s steam tractor and moldboard plow that could break up vast grassland areas. These developments, along with the innovations heralded by Walter Prescott Webb—barbed wire, the windmill, and the six-shooter—may have enhanced European-based optimism in settling semiarid regions for industrial exploitation, farming, or fenced ranching (Webb 1931).[5] The concomitant development of capitalism implied new land-use patterns and new assumptions about land collateral with capitalism's general tendency to individualize ownership, to aggregate individual interest independent of customary communities so as to create investment capital, and to separate ownership from labor while substituting capital for labor (Strange 1988). In areas colonized by English-speaking peoples, settlement and industrial agriculture also meant the destruction of the commons that Hardin finds so insidious and the creation of private property divided along fenced, surveyed lines.

The second enclosure movement in the eighteenth century had fenced or hedged virtually all of the common pasture and woodland in the British Isles and eliminated the open field system whereby medieval and Renaissance farmers held arable lands in strips distributed across the manor that required community cooperation in performing routine farming tasks and a collectively administered attention to the uses and abuse of land (Homans 1960). As the manorial system broke down, constraints also disappeared, and the abuses of the commons that Lloyd described in the nineteenth century grew apace. About 7 million acres were enclosed between 1700 and 1845, after which the General Enclosure Act was passed (Slater 1913, lxxii). But fencing, as John Bennett points out in his essay, made a flexible response to the environment more difficult as the uses of land were determined by fence lines rather than by soil, microclimate, or ecosystem.[6] It also encouraged the use of fragile land. In England "soil that had been reclaimed from the sea, fen land and wild unbroken land that had pretty much to be made over," much of it "relatively poor," was enclosed (Gras 1946, 170–71), and in the Anglophone empire development was also often equally insensitive, because the same system was generally used in most places. Combined with the system of free proprietorship that lifted from the proprietor traditional claims of stewardship and manorial obligation, by 1850 agriculture had come to be, in England and America but not in Russia, a "capitalistic" enterprise (Gras 1946, 259–67). It was to be so also in the semiarid colonies that Anglophone peoples settled.

It may be that indigenes living in semiarid regions experienced less the tragedy of the commons than the tragic destruction of cultures that made the commons possible.

The English humid-lands system was extended to the "dry colonies" under study despite its inappropriateness for the new lands. Though the manorial system was vastly different from tribal systems in most parts of the world, it did, like them, emphasize community enterprise and environmental constraint. The new system did not. Indigenous peoples in the areas conquered were quite often encouraged to abandon pastoralism and hunting and gathering even on their reserves and to adopt European ranching and farming techniques. Sometimes they were rather successful at first, as Barsh demonstrates in treating the Plains Indians and Silitshena the groups in southern Africa (the Kazakh and Aboriginal experience of farming and fenced ranching appears to have been less successful). However, eventually the restrictions placed on them by the allotment and alienation of their lands, the lack of available credit and the inefficacy of European-style farming and animal husbandry on individual plots in semiarid regions meant short-lived success and increasing dependency.

Pastoral nomadic and hunter-gatherer peoples in dryland areas were not always easy targets for the expansive designs of imperial powers; in many cases, for example, the Mongolian khanates of the thirteenth century and after, they were the imperial powers themselves (Lattimore 1951, 5II–49). But industrialism and industrial empire changed everything in the relation between "nomad" and "empire." The theme of the book is as old as civilization itself, since at a figurative level it pictures episodes in the industrial phase of the battle between Nimrod and Abraham, city dweller and pastoralist/hunter-gatherer. Genesis speaks of Nimrod as a "mighty hunter" who switched his occupation to found Babel, out of which came a series of other cities including Ninevah. Immediately after, it speaks of a pastoralist Abraham who wanders with his flocks and tents between Ur of the Chaldees and Egypt but whose heirs are destined to inherit pastoral and agricultural land. But in the biblical account the descendants of Nimrod and Abraham remain rivals, and much of the Hebrew Bible tells of the movement of Israel between a captivity to empire and the freedom of self-determination, often symbolized in the Psalms and the prophetic books by the freedom of the pastoral life. It was possible in biblical times to romanticize the pastoral life as preferable to urban empire and agriculture—as in some ways stronger. And if one goes behind the biblical *mythos* to archaeology and prehistory, the processes of hostile and interdependent interaction between empire and pastoral/hunter-gatherer peoples appear also to be very ancient. But as Anatoly Khazanov suggests in his essay on the pastoralists of Central Asia (chap. 2), the relationship between the settled state, or empire, and semiarid roving peoples changes markedly with the rise of industrialism, particularly with the manufacture of rifles and other weapons accurate at long distances.

As Khazanov points out, whereas before the industrial revolution indigenous nomadic or seminomadic peoples could use their superior horseman-ship and strategy to invade and annex (or loot) adjacent sedentarized "tradi-tional" empires, they could not do so once industrialized imperial warfare appeared, supported by the economic power of Britain and the United States and to some degree the lesser strength of Russia in the nineteenth century (Khazanov, this volume; cf. Khazanov 1978). Industrialized warfare permitted full conquest and annexation of the territories of the peoples dis-cussed in this book, and with the rise of "evangelical" versions of European religion and the pseudogenetics of race, these people were seen not only as enemies to be conquered but as things to be used, "saved," or marginalized in the grand design of empire (Montagu 1965, 23 ff., 44 ff.). They could be treated as people the Great Father had turned his face from—partly because they had few Colt pistols.

At the same time in the eighteenth and early nineteenth centuries as in-dustrialism was making new weaponry available to the European-based em-pires, they were also rapidly expanding their populations. They were buying the idea of progress and following their commercial companies with land armies and colonial administration. Gradually the European empires—par-ticularly the British—moved beyond the coastal areas in Africa, Asia, and North America and annexed the semiarid heartlands. The movement was not capricious to the Europeans. The population of Europe had risen from 86 million in 1700 to 132 million in 1800, to 213 million in 1850 and to 284 mil-lion in 1900 (Clark 1967, 64). England and Wales grew, in approximate fig-ures, from 6 million in 1700 to 9 million in 1800 to 18 million in 1850 to 32 mil-lion in 1900, despite massive out-migration (Grigg 1980, 165). On the margins of Europe, population growth was even more rapid: the semiarid Ukraine tripled in population in the eighteenth century, and the American colonies grew from 350,000 to 5 million in the same period (McNeill 1983, 32–34). Meanwhile in the temperate zones, indigenous non-European peo-ples were suffering from plagues that made their zones appear empty to out-siders and attractive for settlement—for example, in the Russian steppe be-cause of the fourteenth-century bubonic plague (McNeill 1983, 14), on the American Plains because of the sixteenth-century smallpox and other Euro-pean diseases deliberately and accidentally spread.

Like Don Quixote galloping off on his horse, the eagle of capitalistic em-pire traveled from Europe in all directions at once, and it carried with it an agricultural and land tenure system that was new not only to the colonies but to much of Europe. In both England and America, the efforts of liberal reformers to replace traditional communal fields with larger fields that would encourage individualistic entrepreneurial farming, reformed agronomic practices, and modern equipment forced multitudes of rural people out of the countryside and into factory work in the cities. Many had to emigrate to new lands (Mingay 1975, 98–139; Watson 1960, 519–24). The breakup of the old agriculture also meant that "communal" methods of allocating responsibility and fixing jobs according to traditional roles were replaced by private entrepreneurship. Hence, when European-based moralists encountered the role systems of "tribal" peoples, they saw them as immoral or impractical and endeavored to destroy them through allotment and similar devices described in this book (Homans 1960, 109–329; Orwin and Orwin 1967, 106–60; Debo 1970, 251–67).

Thomas Jefferson Morgan, commissioner of Indian affairs in the United States from 1889 to 1893, thought he was destroying socialism and encouraging "respect for individual rights" by allotting land and imposing on Indians "the universal custom among the civilized people of the country" (Morgan 1892a, 24–25; 1892b, 17); not surprisingly, at about the same time in Botswana, John Mackenzie, a leader in colonizing that country, was speaking of doing away with the communistic relations of the members of a tribe and introducing "healthy individual competition" (McCay and Acheson 1987, 179). The histories of the Maasai and the Alaskan Inuit suggest that Western private-property systems are now being imposed on them, more subtly than but just as assuredly as they were imposed on the Plains Indians, many Australian Aborigines, and many southern African groups. At least some of the Plains Indians knew that the allotted property system was not appropriate for running cattle in semiarid lands. Wooden Leg of the Northern Cheyennes called the "little pieces" allotted "big enough for grazing rabbits, but not cattle" (Debo 1970, 266). It would be useful to develop a collection of parallel remarks about the moral superiority of free proprietorship and the socialism of tribal ways from other apologists for conquest and to gather indigenous remarks parallel to Wooden Leg's on the problems of semiarid land allotment.

To allot the new lands under systems of free proprietorship, Europeans and European-based peoples had first to conquer and annex them by direct government action or by action of the companies such as the British South Africa Company; and they did so primarily in the period 1830–90 (map 1.3). As D. K. Fieldhouse has argued, an economic factor was occasionally important in the European-based annexation of the lands belonging to the peoples studied here, and the acquisition of new land was sometimes also important, but neither was the primary motive. Rather, annexation generally came when *economic exploitation* (e.g., by the companies) *had become politicized*, and taking over seemed to the European or American governments to be required to continue the exploitive process (Fieldhouse 1973, 459–771). After annexation, early land decisions in the Anglophone world placed indigenous people on reservations so that "white people" could control the greater share of the land outside the reserves in free proprietorship.

After the best land had been occupied, the primary tools for getting local peoples out of the way permanently without exterminating all of them was, in the English-speaking world, the creation of reservations or bantustans. The English had developed reserves for the Irish and Scottish tribes during the Middle Ages (MacLeod 1928), and they early exported this practice to the North American colonies (Kawashima 1969, 1974). After America separated from the British Crown, John Marshall's Supreme Court decisions in the Cherokee cases (*Worcester vs. Georgia, Cherokee Nation vs. Georgia*) defined Indian nations as internal sovereignties, and only with *Lone Wolf vs. Hitchcock* in 1903 was Marshall's decision significantly altered. Between Marshall's time and the *Lone Wolf* decision, the legal basis for the modern American reservation system evolved, though many "tribes" and traditional head chiefs never accepted the changes The reservation system was sufficiently unfamiliar to be attributed only to the United States and Canada by the *Oxford English Dictionary* in 1928 and by the eleventh edition of the *Encyclopaedia Britannica* in 1911.

However, equivalents to the system were developed in the British Empire and Commonwealth in the bantustans created by South African laws passed in 1913, 1936, and 1970 (Davies et al. 1984, 169–70; Butler et al. 1977, 7–12, 35–36), the Maasai reserve created in Kenya in 1940 by the British colonial administration, and the Aboriginal reserves and base camps developed in Commonwealth Australia after the Second World War. In the United States, the Alaskan Native Corporations set up after the passage of the Alaska Native Claims Settlement Act in 1971 were a kind of alternative to the

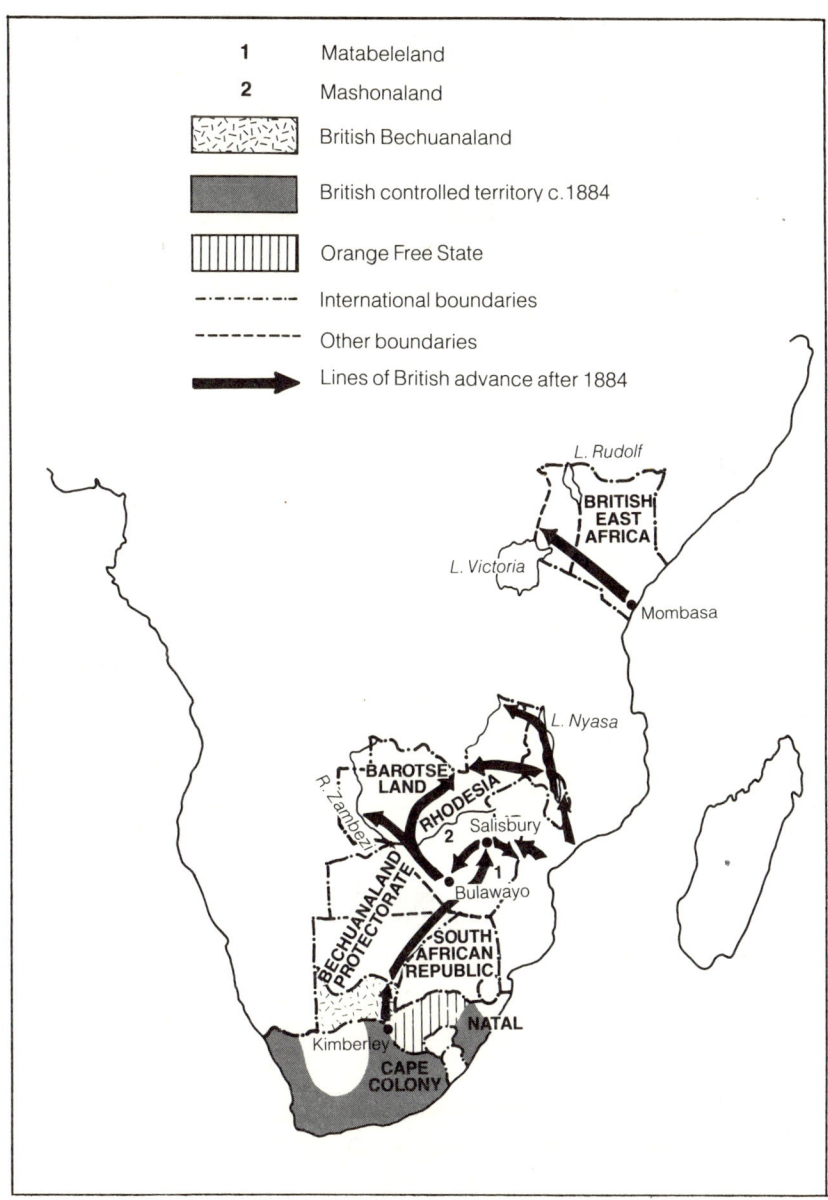

1   Matabeleland

2   Mashonaland

British Bechuanaland

British controlled territory c.1884

Orange Free State

International boundaries

Other boundaries

Lines of British advance after 1884

Map 1.3   British movement into southern and eastern Africa.

reservation. When the reserve no longer contained a people regarded as internally sovereign, new institutions for extending colonialism into the reserve were easy enough for the national governments to create. Though these institutions have had some differences from region to region, depending on historical traditions, they have basically functioned to confine previously nomadic people to small areas—setting them in one place, forcing new economic institutions on them, and serving to alienate their corporate land.[7] Later decisions often altered indigenous land tenure systems within the reserves, allotting community lands to free proprietorship and eventually selling off the land (or taking it away through other means), often forcing indigenes into a condition of dependency. These processes—described in the essays of Barsh, Anders, Bekure and Pasha, Silitshena, and Hamilton—required a continuation of the colonial annexation effort *into the reserve*. That is, reservation neocolonialism only continues what colonialism began.

It should not be thought that this Anglo-American pattern is universal. It has been claimed that the Spanish empire adopted a different system for making servants and serfs of the indigenes, one that led to more internal indigenous self-governance and a higher incidence of marriage with members of the dominant culture (Spicer 1969, 24–48; Haring 1947, 42–75; de Madariaga 1947, 248–65). Yet whatever its virtues, the tradition of the Spanish imperial system did little to spare Indians of the semiarid pampas once the Spanish empire had been overthrown and Argentina established. General Julio A. Roca began his campaign against the Araucanian Indian pastoralists in the semiarid pampas adjacent to the Río Negro—the last Indian resistance in the country—in 1879–80, about the time of the last military campaigns against the Plains Indians (Wright et al. 1978, 44, 204, 412–14). He butchered them like cattle, leaving only a few isolates to flee and mingle with other populations (Rennie 1945, 121–27). Since the Araucanians had by that time absorbed the other pampas groups, this book can contain no section on the indigenous peoples of the pampas, all of whom were killed or forced to intermingle instead of being colonized.

In contrast to the Argentine practices, the Soviet Union rejected both reservations and genocide in the formerly pastoral regions of Central Asia and chose a "peoples" policy of establishing "native soviets" and later national districts where minorities could preserve some aspects of their culture while they were being assimilated (Spicer 1969, 28–48). At the same time, immigration into formerly pastoral nomadic territories from European Russia was used to integrate the area into the larger unit (Armstrong 1966; Carrere D'Encausse 1981).[8] Though the process of settling the Central Asian cultures and then assimilating them into the industrial state has proceeded more slowly in Soviet Central Asia, partly because of the force of indigenous numbers and partly because of the deference the Soviets have paid to the separate cultures in their territory (Pipes 1964), the process has been steady and fairly draconian, as Khazanov's essay attests.

The consequences of industrialization and industrial agriculture for the lands and peoples discussed in this book were disastrous, often because of deliberate European-based exploitation but also because of Europeans' inexperience in the dry worlds.[9] The ancestors of the European empires grew up in ancient times in dry Near Eastern river basins (Hills, 1986), and much of the land surrounding the Mediterranean was semiarid country, but the great nineteenth-century colonial empires had forgotten the lessons of such places and expanded into the semiarid regions largely from humid and sub-humid climates (McNeill 1964).[10]

Europeans *had* experience of semiarid regions outside Europe before the 1830s: the Spanish had successfully exported the *vaquero* culture to Mexico, where it may have formed the basis for open-range ranching in Mexico and on the Great Plains (even Terry Jordan does not question the dominance of the Hispanic influence in the creation of Texas cattle culture, though he argues for a secondary southern influence; Jordan 1966; Bishko 1952, 491–515, 1963, 47–69); the Boers had developed a European semiarid pastoralism in the Orange Free State and the Transvaal (Marks and Atmore 1980, 1–43; Marks and Rathbone 1982, 1–43; Trapido 1978, 26–58); and Russia had experimented with semiarid agriculture on her southeastern boundaries (Matley 1967, 266–308).[11] But the dominant nation-state in each of these cases was not industrialized (or was only slightly so, in the case of Russia).

The European grand entrance with industry and industrialized agriculture to the semiarid regions under consideration came after 1830: sedentarized settlement came slowly to the Kazakh regions of Central Asia after the 1830–54 Russian conquest (map 1.4), to the Great Plains in 1854–90, to central Australia in the 1860s–70s, to much of Kenya after the period 1884–94 when British East Africa was created, and to the Matabele and Shona lands in Africa—given to Cecil Rhodes as Rhodesia—after 1889 (Boer settlement

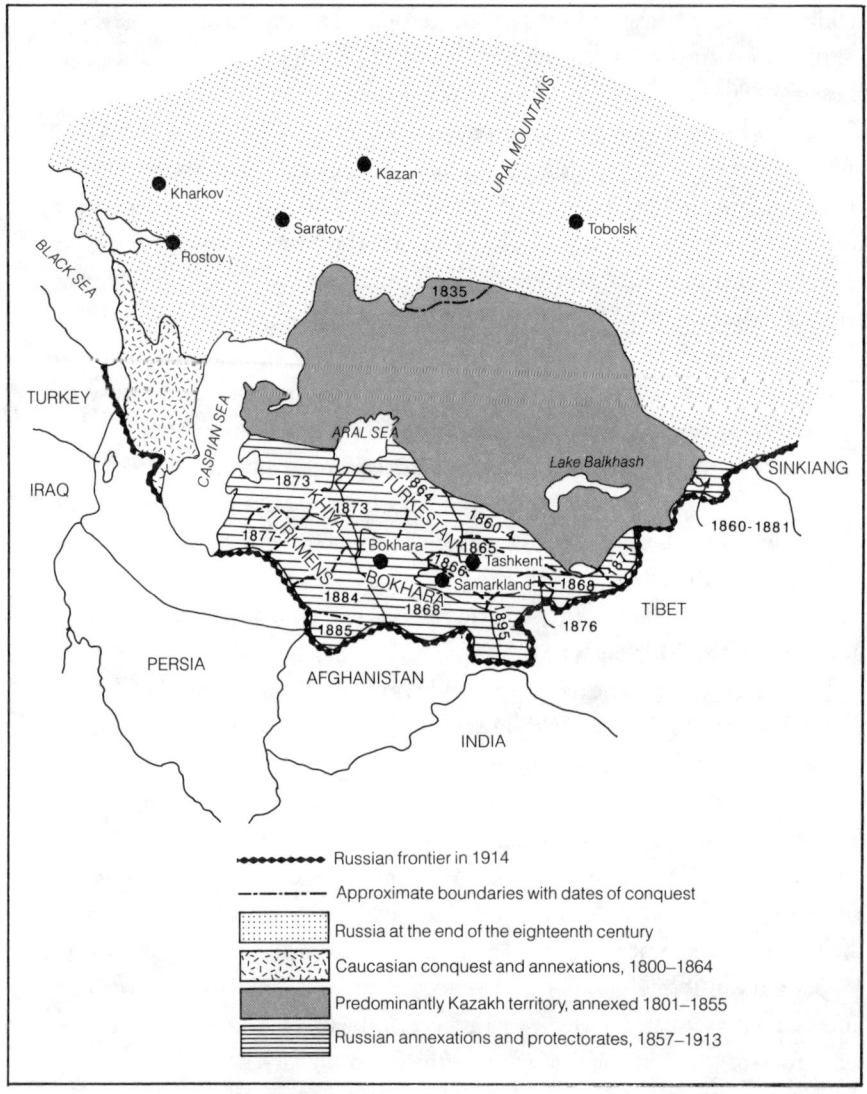

Map 1.4 Russian conquest of Kazakhstan and Central Asian lands.

had come earlier, but it was not industrialized; map 1.4). The arctic regions of the Inuit and Lapps (Saami) were heavily affected by industrial culture only after World War II, when techniques for using the energy available in the Arctic came of age.

At least some of the indigenous peoples treated in this volume recognized the potential inefficacy of the resource-use regimes forced on their former lands and had their own versions of Washakie's explosion. For example, Black Elk of the Oglala band of the Lakota Sioux, discussed by Callicott, believed that everything in the far-horizoned world through which his people searched for bison moved in a circle like that described by his culture's first finders of the four directions. The circle was the model for the sky, the earth, the realm of the stars, the wind in its whirling, the movement of the sun and the moon and of the seasons and the ages of man, and if formed the pattern for both the nests of birds and the tents of human beings (Neihardt 1961, 198–99).[12] Annette Hamilton also speaks, in her essay, of this cyclical sense as being basic in the perception of the Australian Aborigines.

The civilization of the Europeans appeared as a stark contrast to Black Elk. Based on linear notions of postenclosure surveyed private property and straightforward progress, it seemed to segment everything: "square boxes" for houses (Neihardt 1961, 198, 217–18), the lines of reservations to "keep [the people] in" (218), the "grass penned up" (221), and prisons that "made [people] move around like animals in a cage" (221). The conquerors who drew these lines did "not care for each other," and people were willing to "take everything from each other if they could" (221). It appeared that the people who imposed the new regime had "forgotten that the earth was their mother" (221).[13] Though Black Elk's statements may include both nostalgia and bitterness and certainly derive from the Lakota religious system, they are also accurate statements about contrasting ways of handling the land and its resources, reflecting the difference between the English-American system of private property and traditional Lakota usufruct systems. Similarly, according to Robert Hitchcock, the San in Botswana said that Bushmen lived on wild plants and animals, black people on crops and cattle, and white people on money, bread, and sugar—that to turn land into crops was to deny the San their God-given tradition. Similar statements could be collected from the folklore or belles lettres of many of the non-European peoples represented in this book.

When Black Elk said it appeared to him that the whites had "forgotten that the earth was their mother," he must, given the role of Maka-Earth in Lakota religion, have anticipated that European ways could be self-defeating in semiarid regions, and they often were. In a recent publication, written by a food panel advisory to the World Commission on Food and Development, entitled *Food 2000*, that group indicates that 29 percent of the world's land area, supporting over 230 million people, is subject to what it calls desertification based on "largely man-made processes" (Advisory Panel 1987, 67–68). The report points to areas in North America, the USSR, Africa, and Australia as undergoing this process, including many of the regions formerly inhabited with full sovereign power by the indigenous groups discussed in this book. The Organization for Economic Cooperation and Development (OECD) *State of the Environment: 1985* report shows the Great Plains as at moderate risk of desertification, central Australia and the Kazakh regions of the USSR as at moderate to high risk, the Maasai regions of Kenya as in the moderate risk zone, and the areas of southern Africa once held by the various groups described by Robson Silitshena as in the moderate to high risk zones (OECD 1985, 96).[14] Only the Alaskan regions described by Anders escape in the OECD analysis (map 1.5), and perhaps they should not.

In the process of alteration, the land rendered useless for any economic return includes 21 million hectares annually (about 52 billion acres), especially in the developing countries (Advisory Panel 1987, 69). The authors of *Food 2000* attribute the process to drought, but also to human and livestock population growth (p. 72) and to *the development of policies that favor cash over subsistence crops or force herdsmen to raise their animals on unsuitable land*: "In the pursuit of quick gains, urban entrepreneurs have engaged in economic activities without sufficient appreciation of the environment" (Advisory Panel 1987, 71).

Though *Food 2000* suggests that the semiarid regions in the United States and North America are not declining appreciably in quality (p. 68), they nevertheless are endangered by the same problems that affect the other regions studied here (Rosenberg 1986). The areas of North America once controlled by the Plains Indians have not had a visible environmental disaster of the "dust bowl" variety since the 1930s, but minibowls have been common. Soil erosion through washing removes five tons of topsoil annually for each acre of farmland (Worster 1984, 56–67). Some observers argue that

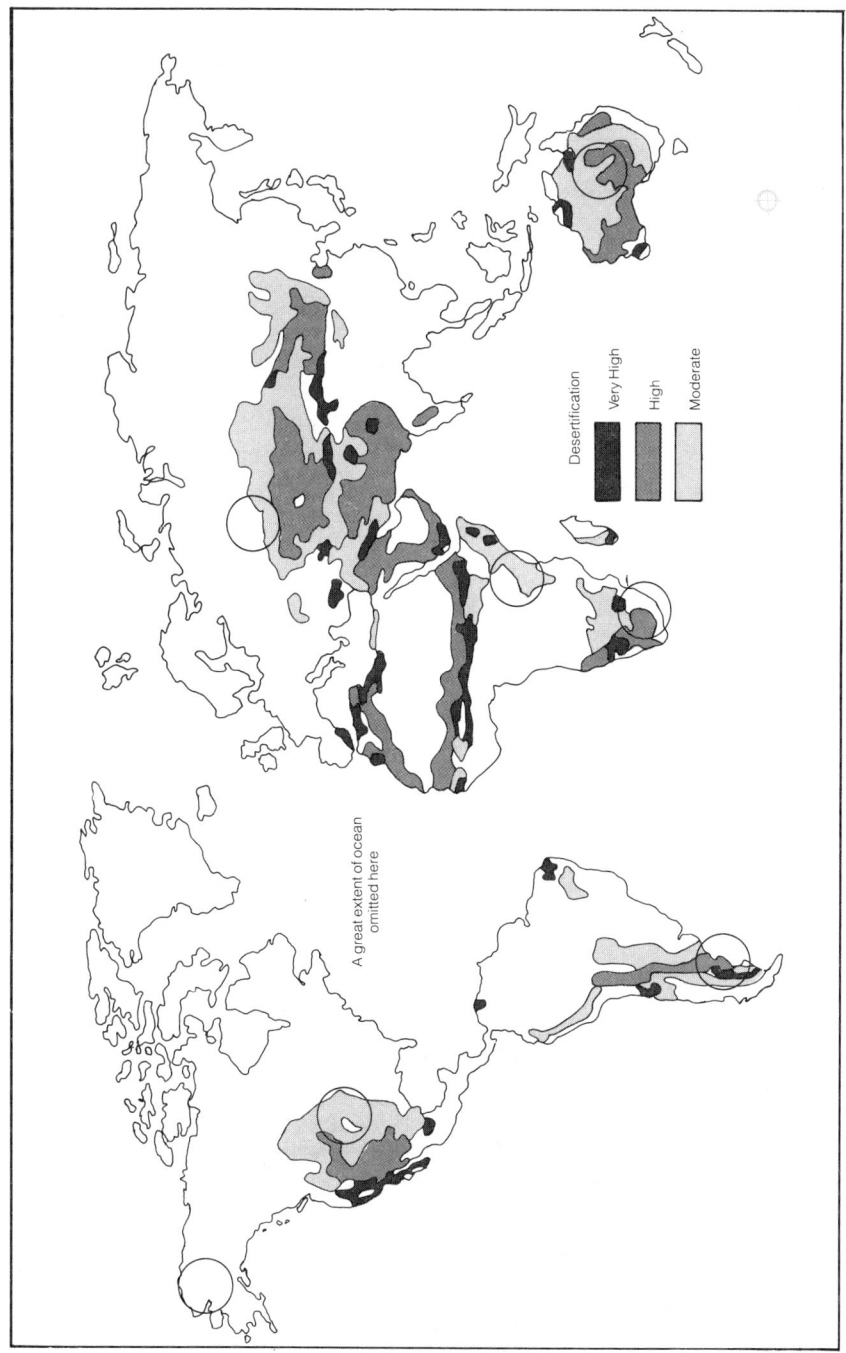

A great extent of ocean omitted here

Desertification
Very High
High
Moderate

Map 1.5 Worldwide desertification in semiarid regions.

the topsoils of the Great Plains hardly exist anymore. Over 60 percent of the land in the Great Plains corn belt area has experienced annual water erosion of more than ten tons per acre per year (USDA 1980, 50, 82); Lewis and Reinsch found that soil erosion in eastern Nebraska exceeds the depth of the original topsoil and that 10 percent of western Nebraska has become unproductive because of wind damage (1984, 17). Donald Worster's hypothesis that overly intense development pressures caused the Great Plains thirties dust bowl suggests that Great Plains development policies have been identical with those that created the recent disasters in the Third World (Worster 1979). That the Great Plains problems were the result of internal rather than international development does not make them different in kind, since much the same market mechanisms operate in both contexts.

These major developmental failures in the semiarid world have led international agencies to emphasize restudying small group methods. For example, the Swaziland Manpower Development Project of the United States Agency for International Development built its economic development activities on the traditional woman's group practice of holding work parties to strengthen the women's groups *and* assist them to develop leadership and income. Robert W. Chambers has argued, in a paper written for the World Bank, that if the bank wishes its environmental policy to succeed, it ought to invest in poor people who have long inhabited a region and have an investment in the environment's protection for their survival—a policy that, if implemented, would often mean fostering extensions of the traditional productive efforts of peasant and tribal small groups (Chambers 1987), and other development economists associated with the World Bank are rethinking single allotment as the ideal land reform (Falloux 1988, 190). The full body of the World Commission on Environment and Development, in *Our Common Future*, argues that though the isolation of small-group societies has often meant their marginalization, dispossession, and even cultural extinction, it has also often meant "the preservation of a traditional way of life in close harmony with the natural environment" (World Commission 1987, 114). Many anthropologists, including some of the authors in this book, might question the appropriateness of the word "harmony," but they would almost certainly endorse the general position that small-group societies have a "vast accumulation of tradition, knowledge and experience" and that their destruction "tends to destroy the only cultures that have proved able to thrive in these environments" (World Commission 1987, 115). Indeed, the case studies in this book exemplify the World Commission's assertion, though they were written before *Our Common Future* was published.

At a pragmatic level, the World Bank's Technical Paper 61, dealing with desertification in the Sahel, speaks of the defects that past industrial development efforts have shared in arid/semiarid regions: "Planners have often misunderstood the logic of traditional production systems, and have thereby overestimated the ease with which improvements could be introduced and underestimated the negative consequences of intended improvements" (Gorse and Steeds 1987, 19), especially through overly optimistic projections on the basis of annual rainfall averages rather than minima and maxima (cf. Baker 1986, 238–45).[15] The essays below, especially those of Bennett, Khazanov, Hamilton, and Barsh, also suggest that planners may have underestimated the productivity and sustainability of traditional systems.

Until we have a good series of estimates of the likely total food productivity of traditional and industrialized systems over time, operating within approximately the same environments, we cannot make the assumption that "industrial systems will support a larger population." The essays by Barsh, Bennett, and Hamilton suggest that may not always be the case; in the radically different environment of the Lacandon rain forest, James D. Nations and Ronald B. Nigh have shown that traditional production systems produce 1,800 kilograms of corn per hectare, plus squash, sweet potatoes, yams, jicamas, and beans, as well as deer, squirrels, pacas, and peccaries that are allowed to harvest part of the small Lacandon fields and are in turn harvested. In addition, the milpa produces fruit and root crops during fallow. This system, a modification of the ancient Mayan system, contrasts in its permanence and productivity with the commercial livestock production systems of the area, concentrating on beef products that yield limited amounts of food and erode the soil irretrievably within decades. Nations and Nigh suggest that the "food production system of the Lacandon Maya offers a rational base for the development . . . which could produce a healthy diet as well as export items for farming communities" (Nations and Nigh 1980, 8–19, 26).

At the other end of the rainfall scale, John Yellen and Richard Lee's examination of the Dobe/Du/da Kalahari indigenes of the Dobe/Du/da area indicates that the foods found in that semidesert are "so plentiful for most of the year that the !Kung can afford to exercise selectivity in their diet" and that such a system, left alone, could continue indefinitely so long as the "natural vegetation cover" is retained (Yellen and Lee 1976, 37–42, 46). Yellen and Lee can give no precise production figures per area of land; Russel Barsh, in his essay in this book, gives estimates for bison productivity on the Great Plains. However, until studies like Nations and Nigh's are replicated in a large number of regions, including prototypical semiarid ones, we will not be clear about the food production consequences over time of traditional and industrial systems. What *Food 2000* makes clear is that industrialized production systems are not sustainable in their present form in semiarid regions.

Why did industrial civilizations often "fail" to create sustainable systems in the regions under study while indigenes "succeeded"? First of all, we should not assume that the indigenes always succeeded. For example, the Bangwato of Botswana had to move their capital three times because the area around it became degraded.[16] Concomitantly, Bennett suggests in his essay that European-based peoples may "succeed" when not drawn too fully into the centralized systems. Degradation has always existed, but not on the modern scale. In particular, the groups studied in this book *have* generally maintained a sustainable relationship to their surroundings and produced "enough" before European intervention in their affairs.

Some reasons for semiarid environmental degradation under industrial systems are rather obvious. Hunter-gatherers and pastoral nomadic peoples, even those who practice some agriculture, leave the floral cover intact except in alluvial soil along rivers and creeks, where they may practice agriculture; the mobility of their resources means they can respond readily to initial signs of the degradation of plant cover. This argument, represented more fully in the essays by Bennett, Barsh, Hamilton, and Khazanov, suggests that mobile, nomadic ways of getting resources from land *work* partly because animals and people move when forage is short. They have to adapt to the immediate and respect the homeostatic controls nature places on them, including controls on births and deaths.[17]

Second, what is used for food is adapted to the region and does not require heavy transformation of soil, water supply, terrain, or other factors through energy output. Bison, as Barsh points out in his essay, are adapted to the Plains; the animals and plants the Australian Aborigines use for food belong to their terrain, and so forth.

Third, indigenous regimens do not subject fragile lands to the destructive effects of the machines necessary to modern industry—automobiles (Hamilton), mining equipment (Hamilton), dams (Morris), massive plows (Khazanov, Silitshena), or oil extraction and piping devices (Anders). This destruction, indiscriminately perpetuated by both indigenous and European-based people once the machinery has arrived, can be and has been consciously rejected by some people with access to it, such as the Amish, some Australian Aborigines, and even North American Indians in more remote areas (Hamilton, Callicott). But the temptation of industrial ease is difficult to resist.

Fourth, indigenous peoples may have consciously chosen to respect the plant and animal communities in their surroundings and erected their religious and ethical systems on this choice. If societies now living in such regions are to change directions and develop a more parsimonious regime, they will have to do so *consciously*. Past regimes will not spontaneously reconstruct themselves, and the momentum of the world economy encourages development that leads to global disequilibrium. That conscious choice of future directions is possible in drylands may be indicated, at least superficially, in Annette Hamilton's account of Aborigines who return to their base camps to pursue a traditional life or by such groups as the Amish. O. Douglas Schwarz argues that the conscious choice of a conservation ethic, based on that of the Plains Indian societies, is possible to modern industrialized people and is being practiced by some of them, and Baird Callicott argues that some Plains cultures did possess a land ethic that we could emulate.[18]

But not everyone believes such choice is possible or based on belief. Schwarz's and Callicott's arguments contrast with those of other scholars who have argued recently that there is little detailed evidence that indigenous people actually "conserve." What appears as conservation may be necessity. One such scholar, Raymond Hames (1987), argues that to show that ideology affects treatment of the environment, one would have to demonstrate that actions were generated by beliefs, were designed to solve an environmental problem, and did in fact quantitatively affect the surrounding world so as to "conserve" a cherished resource. The territory where such actions occurred would have to be rich in resources and its people capable of defending it against outsiders who might wish to upset its conservation plans. Further, the group would have to have mechanisms for punishing polluters and wasters and have no notion that areas other than the home territory were available for pursuing the food search (McCay and Acheson 1987, 92–107). Hames's argument is a useful antidote to easy romanticism about indigenes, and given his stringent criteria, conservation cannot be demonstrated to have been practiced by the peoples studied in this book. Indeed, conservation probably cannot be shown to have existed in any semiarid region, since the peoples indigenous to these areas have not been able to defend their resources for some time. Even in precolonial times, the claims of nomadic and seminomadic peoples on territorial resources often overlapped.

However, though one may not be able to demonstrate *conscious conservation* in semiarid regions, one can show a congruence between belief (ethical system, religion) and practice that is the equivalent of a developmental choice. For example, Strehlow (1965) argues, and Hamilton agrees, that the Aboriginal concept of sacred places as regenerative ensured the survival of species in times of drought because the belief system forbade taking water or game from such places. Ross (1978) shows that food taboos in the Amazon basin apparently help in the conservation of species necessary to the total ecosystem and to human survival (cf. Vickers 1988). The Omaha ritual of the buffalo hunt celebrated immediately after the kill gave clear directions on how the buffalo was to be butchered and conserved that were paralleled in the actions of the hunt (Olson 1979, 27–45; Fletcher and La Flesche 1972, 1: 275–309, esp. 307).

Though demonstrating consonance between belief and practice may be possible, what may be crucial to our understanding is not any one facet of an indigenous group's life on the grasslands but its total set of adaptive assumptions and practices compared with the industrial set, each set conceived as a system. The differences undoubtedly would be related to the problems that industrialized societies have encountered in semiarid zones: that is, industrial societies in semiarid regions may not be "adapting" to soil, grass, animal, and weather so much as they are responding to the demands of the centralized power that invested in the development of industry and industrial agriculture in the area to obtain returns characteristic of less fragile, more resource-rich areas.

This view is described more fully in John Bennett's *The Ecological Transi-*

tion. There Bennett, setting forth many of the assumptions that underlie his essay in this volume, argues that one can "reduce the complex and gradual changes of the historical process to two [ideal] types: societies that find an equilibrium with Nature and those that do not" (Bennett 1976, 136). He further argues that one finds in the latter type of society a "progressive incorporation of Nature into human frames of purpose" (1976, 3), such an incorporation as, for example, accompanied the enclosure movement in England and its colonial offspring. In Bennett's view, equilibrium societies do not exist in some mysterious "mystical harmony" with nature but fluctuate between "overuse and regeneration, on a homeostatic basis," seeking, however, to achieve a "sustained-yield system of resource use" (1976, 137). They succeed in doing so because of their structure and their relation to the natural world. Bennett represents the feedback mechanisms that permit equilibrium societies to function in effective relationship with the environment and prevent industrial societies from doing so as shown in table 1.1.

As the table argues, equilibrium societies tend to be small, place a high percentage of their citizens in direct contact with the environment, and hence get feedback from it.[19] By minimizing the needs of their citizens, keeping their expectations low, and developing simple technologies and a cultural ecology designed to control resource use, they avoid degrading their surroundings. In contrast societies in disequilibrium tend to be large and provide little environmental contact for most of their citizens. They transfer resources easily from region to region, maximize sustenance needs, and expand expectations for gratification while employing a complex technology that promotes resource use. Obviously, training people to accept controls, use local resources, and minimize needs is a matter of direct instruction in behavior as well as of general mythos.

If one looks at the histories of the Great Plains, central Australia, Kazakhstan, southern Africa, Kenya, and the Inuit and Saami areas described in this book, the movement of the societies in these areas under industrial impact has been away from immediate contact with the environment, away from parsimony, away from using primarily the resources available in the local area, and toward agricultural, hydroelectric, and mining activities that require extensive external inputs and produce outputs that go to external consumers. Scholars have criticized Bennett for failing to include the built environment in his systems model (Choldin 1979) and for

Table 1.1 Feedback Mechanisms Influencing Equilibrium

	Societies in Equilibrium with Environment	Societies in Disequilibrium with Environment
Population dynamics	Small, controlled	Large, expanding, weakly controlled
Contact with environment	Direct contact by maximum number of people	Direct contact by minimal number of people
Range	Restricted to local resources	Resources available from external sources
Sustenance needs	Close to minimal; defined largely by physiological needs	Maximal; defined in large part by cultural wants
Gratification expectations	Low; controlled	High; promise of continued expansion
Technological capacity	Low	High
[Consumption system]	Functioning to control resource use	Functioning only to promote resource use

assuming that "cultural motivations alone" are responsible for environmental problems (Hardesty 1977), and these criticisms may be accurate. However, his general approach has been used with a large number of ecosystems and continues to form the basis for much analysis of the relation between culture and the environment. It gives at least one explanation of the group of phenomena under consideration here. Bennett's argument points toward decentralization and provisions for clearer feedback from nature as well as belief system support for parsimony and flexibility.

If such assumptions as Bennett describes are to inform present choice, as suggested by *Food 2000*, they must be protected, studied, and tried. They can be so, considering the millennia of indigenous experience now being destroyed, only in a framework of international and national law that guarantees that resources will not be alienated from the group through industrial development outside its control or through "allotment" that replaces the group with the individual (for discussions of this issue, see the essays below by Morris, Hamilton, Silitshena, Bekure and Pasha, Barsh, and Anders).

Meanwhile, we do have some stopgap measures. A number of efforts to stop the degradation of semiarid lands consciously or unconsciously follow the principles of indigenous regimens. For example, no-till agriculture, in leaving as much cover and root structure in the soil as possible, follows a principle of much "tribal" agriculture. More pronounced, the experiments with perennial covers of Wes Jackson at the Land Institute in Salina, Kansas, avowedly are inspired in part by Plains Indian hunter-gatherer practice; those of William Mollison in Australia are likewise based on aboriginal practice (Jackson et al. 1984, xiv and passim; Mollison 1984, 1986). So also is the Savoury system of short-duration grazing in Zimbabwe. Jackson has further argued that "civilized agriculture" has been wrong from the beginning in relying on annual monocultures (Jackson et al. 1980, 1984; cf. Popper and Popper 1987). Turning to drylands indigenous agriculture, Gary Nabhan's *Gathering the Desert* shows that the Indian tribes of the Sonoran desert find uses for more than 425 edible plant species that can be gathered or fostered there, including the cresote bush, agave, organ-pipe cactus, *Amaranthus palmieri*, the tepary bean, bird pepper, buffalo gourd, and coyote gourd. Each is dependent on an unaltered climate and a network of often endangered plants and animals for its survival, but each survives in a desert climate without irrigation or the addition of farm chemicals. Nabhan has explored the possibility of a desert agriculture based on this southwestern practice (Nabhan 1986) and is himself developing seeds from native sowers in the Southwest and beginning a similar search among Plains agricultural tribes.[20]

On the Great Plains, one finds extensive experimentation with bison instead of cattle because they are more efficient grazers and better adapted to the harsh Plains climate. In Kenya the savanna eland, once the target of eastern African indigenous hunting, has been used successfully in domestication experiments because it is better adapted to the savanna and more resistant to African diseases than European cattle breeds (Harris 1980, 376–77). According to Robert Hitchcock, the Tswana of Botswana are using traditional land divisions—into grazing areas and arable—as the basis of modern determinations of land use; pastoralists in Syria are making similar determinations based on traditional grazing divisions. In the wake of the Plains Indian civil rights movement, some Plains tribes have experimented

with bison herds and other cooperative enterprises based on traditional social organization and relied more on traditional religious and head chief
structures for leadership. Following up on this movement, about a decade
ago, David Brokensha produced a volume concerned with the practicality
of attending to indigenous developmental practices, including those in
semiarid regions (Brokensha et al. 1980). The existence of these experiments
and the success of some of them suggests that future development in lands at
risk may require modifying of industrial processes, a development from indigenous practices, or a combination of both (Little 1987). All these efforts
assume that human beings *can* plan and choose somewhat rationally in matters involving values, resources, and development. If they cannot, the issues
presented here are beyond consideration.

One should note that, despite the oppression and exploitation, all may
not be lost. Many Aboriginal and Plains Indian peoples have been able to
convert their reservations from "prisons" to "homes" (Iverson, this volume); despite their problems, the Maasai have been able to make a go of
some of the group ranches and retain a group identity; some of the Alaskan
native corporations are working; the peoples of Botswana, though undergoing tremendous droughts, are retaining some of their tradition and "conservationist" practice; and the Kazakh of the Soviet Union are at least reas-
serting their identity as a people. In many lands, traditional group religions
are on the rise, and at least the beginning of concern for international legal
protection for indigenous people is developing, as exemplified by the
United Nations Declaration and Program of Action to Combat Racism and
Racial Discrimination (1978) and the International Labour Organization
1957 Convention 107 ("Concerning the Protection and Integration of Indigenous and Other Tribal and Semi-tribal Populations in Independent Countries").[21] In danger of disappearing are many of their practices, attitudes,
and beliefs—wisdom that could be of use to the human race in general.
Those of us who live in dryland areas may all have to "damn a potato" before
we complete the saga of desertification and degradation of our lands.

THE STRUCTURE OF THIS BOOK

Part 1 explains the *history of human adaptation* in semiarid lands. The first essay, by John Bennett, is based on his explorations in systems and decision

theory, and the second essay, by Anatoly Khazanov, is based on Khazanov's own refinement of the approaches to historical anthropology developed in the Soviet Union. Though Bennett and Khazanov begin with quite different approaches and disagree on whether the mounted Plains Indians should be considered pastoral nomads or hunter-gatherers, their conclusions about the environmental effects of the intervention of industrial cultures in semiarid regions are similar and their descriptions of the adaptive process remarkably alike.

Part 2, containing essays by Russel Barsh, Gary Anders, Robson Silitshena, and Peter Iverson, deals with the *history of land use* in semiarid regions. It shows the process by which indigenous peoples were sedentarized and their land regimes replaced by Western ones, and it explores the effects of this replacement on indigenous people and their resources in the Great Plains (Barsh), Alaska (Anders), southern Africa (Silitshena), and Australia (Iverson).

Part 3, containing essays by Patrick Morris, Annette Hamilton, and Solomon Bekure and Ishmael Ole Pasha, moves from land to *institutions for exploiting the land and its resources*—institutions created to "use" the resources of indigenous people and at the same time integrate them to market and world systems in a colonialized or "neocolonialized" way. The institutions considered are the reservation, the internal resource agency ("Department of Interior," Morris), the base camp in Australia (Hamilton), and the group ranch in Kenya (Bekure and Pasha). One may add to this the Alaskan native corporation described by Anders in the preceding section.

Finally, part 4—by O. Douglas Schwarz and J. Baird Callicott—looks at the extent to which indigenous religions may be characterized as conservationist and how far they may influence belief and action in the industrialized world. This section uses only Plains Indian religions as examples, but one would expect its methodology could be more generally applied and developed in conjunction with more detailed behavioral analyses in areas where the group beliefs and practices are still sufficiently integral to one another to make such a project feasible. In each of the four parts, an essay dealing with Great Plains issues comes first, followed by essays on related issues in other regions.

NOTES

1. Cf. Virginia Irving Armstrong (1971, 128) and Oliver LaFarge (1956, 191). LaFarge attributes the remark to the Smithsonian files. As the author of the Introduction, I claim no special scholarly expertise in developmental issues. I have specialized in American Indian literature as one of my interests and have consulted for twenty years with some of the Plains tribes in matters of culture and economic development as well as with the Lincoln, Nebraska, Indian Center.

2. For a useful set of essays discussing land and drought problems of semiarid lands in North America, Central Asia, Australia, and Africa, see the "International Drought Mitigation" issue of the *Great Plains Quarterly* (vol. 6).

3. For an account of the history of environmentalist and culturalist explanations of hunting in one semiarid region, see Blouet and Luebke (1977, ix–xxvii) and Bennett (1969). This book does not make much of the differentiation of the various European ethnic adaptations to semiarid regions and to indigenes but rather emphasizes the common procedures for handling such lands developed by nineteenth-century industrial peoples. We recognize that European ethnic differentiations are also important, but they are for the most part beyond the scope of this volume.

4. The significance of the invention of breech-loading light arms accurate at a distance in the conquest of highly mobile semiarid peoples has been underestimated. For some assessment of the African situation, see Headrick (1981, 83–128).

5. Industrialization depended on a faith in "progress," the faith that nature could be remade to conform more closely to human desire and need. For the evolution of this idea, see Bacon (1924, 35), Jones (1951, 10–40), and Black (1970). For further discussions of the implications of the idea of progress for semiarid indigenes, see Hamilton, this volume, and the introduction to part 4.

6. For a good analysis of the importance of flexibility in the use of semiarid regions, see Fortmann (1986).

7. For a good account of the processes of creating semiarid dependency, see White (1983).

8. For a review of the difficulties European-based peoples had in adapting to one grassland environment, see Bogue (1981, 105–31).

9. For a further analysis of Soviet substitutes for an assimilation policy in Central Asia, see Bennigsen and Wimbush (1978).

10. For a useful account of the details of ancient semiarid farming, see Ellison (1980–83).

11. European settlers began to move into Ukraine and the western parts of the steppe in the sixteenth century.

12. For a transcription of Black Elk's original words on the circle, see DeMallie (1984, 123, 129, 290–91).

13. For a similar transcript, see DeMallie (1984, 245–48, 240, 256, 290).

14. For a survey of worldwide problems in semiarid land use in the interwar years, see Powell (1986). For useful surveys of issues in many of the specific countries treated in this book, see Fortmann (1986) for Botswana, Rosenberg (1986) for the Great Plains, Heathcote (1986) for Australia, Baker (1986) for Africa, and Lydolph (1986) for the Soviet Union. Each of these essays emphasizes a significant role for overdevelopment and the employment of industrial-agricultural technology in inappropriate ways.

15. For another World Bank paper that suggests observing the techniques of environment-management of the poor and rooted peoples of a region as the basis for stabilizing the environment, see Chambers (1987).

16. I obtained this information from Louise Fortmann of the University of California at Berkeley.

17. For a qualification of the concept of homeostatic controls among nomadic peoples, see Khazanov (1981).

18. The issue of indigenous conservationism arises from the concomitant issue of whether Christian or post-Enlightenment Europe possessed a conservationist religion (Berry 1981; Black 1970; White 1983; Martin 1978). For further discussion, see part 4 above.

19. Bennett has written several other pieces that extend the argument of *The Ecological Transition* and relate to the argument of this book (Bennett 1980, 1982, 1986a, 1986b).

20. If the study of indigenous regimes assists the human race in solving the problems described in this essay, it will be the second time tribal peoples have furnished major solutions to humankind. Shortly after Columbus discovered America, New World crops relieved food pressures in a number of Old World regions. However, the land-management systems of New World groups were rarely evaluated or copied (McNeill 1983, 34). For studies of the Indian Great Plains agriculture, see Hurt (1987) and Wilson (1987). Andrew Warren, in an unpublished paper delivered at the 1985 conference of the Center for Great Plains Studies, has argued that much scholarship on pastoral nomadic peoples and the advantages of enclosure is hegemonic (Warren 1985).

21. For useful discussions of indigenous rights, see Gordon Bennett (1978), Foundation for the Study of Plural Societies (1975), and Hitchcock (1985b). The publications of Survival International and the World Council of Indigenous Peoples present current issues.

REFERENCES

Advisory Panel on Food Security, Agriculture, Forestry and Environment to the World Council on Environment and Development. 1987. *Food 2000: Global policies for a sustainable agriculture*. London: Zed Books.

Armstrong, Terence. 1966. The administration of northern peoples: The USSR. In *The Arctic frontier*, ed. R. St. J. Macdonald, 57–88. Toronto: University of Toronto Press.

Armstrong, Virginia Irving. 1971. *I have spoken*. Chicago: Swallow Press.

Arrowsmith, William. 1971. Teaching and the liberal arts: Notes toward an old frontier. In *The liberal arts and teaching education: A confrontation*, 3–26. Lincoln: University of Nebraska Press.

Bacon, Francis. 1924. *New Atlantis*. Ed. A. B. Gough. Oxford: Clarendon Press.

Baker, Randall. 1986. The African experience: Drought and famine in the dry zone. *Great Plains Quarterly* 6:238–45.

Bennett, Gordon. 1978. *Aboriginal rights in international law*. London: Royal Anthropological Institute.

Bennett, John W. 1969. *Northern plainsmen: Adaptive strategy and agrarian life*. Chicago: Aldine.

———. 1976. *The ecological transition: Cultural anthropology and human adaptation*. Elmsford, N.Y.: Pergamon Press.

———. 1980. Human ecology as human behavior. In *Human behavior and environment*, 243–77. New York: Plenum Press.

———. 1982. *Of time and the enterprise: North American family farm management in a context of resource marginality*. Minneapolis: University of Minnesota Press.

———. 1986a. Summary and critique: Interdisciplinary research on people-resources relations. In *National resources and people: Conceptual issues in interdisciplinary research*, 343–72. Boulder, Colo.: Westview Press.

———. 1986b. Research on farmer behavior and social organization. In *New directions for agriculture and agricultural research*, 367–402. London: Rowman and Allanheld.

———. 1988. Anthropology and development: The ambiguous engagement. In *Production and autonomy*. Society for Economic Anthropology Monograph 5. New York: University Press of America.

Bennigsen, Alexander A., and S. Enders Wimbush. 1978. Migration and political control: Soviet Europeans in Soviet Central Asia. In *Human migration: Patterns and policies*, 173–87. Bloomington: Indiana University Press.

Berry, Wendell. 1981. *The gift of good land: Further essays, cultural and agricultural*. San Francisco: North Point Press.

Bishko, C. J. 1952. The peninsular background of Latin American cattle ranching. *Hispanic American Historical Review* 32:491–515.

———. 1963. The Castilian as plainsman: The medieval ranching frontier in La Mancha and Extremadura. In *The New World looks at its history*, ed. A. R. Lewis and T. F. McGann, 47–69. Austin: University of Texas Press.

Black, John. 1970. *The dominion of man: The search for ecological responsibility*. Chicago: Aldine.

Blouet, Brian, and Frederick Luebke. 1977. Introduction. In *The Great Plains: Environment and culture*, ix–xxviii. Lincoln: University of Nebraska Press.

Bogue, Allan G. 1981. The heirs of James C. Malin: A grassland historiography. *Great Plains Quarterly* 1:105–31.

Brokensha, David, et al., eds. 1980. *Indigenous knowledge systems and development*. Washington, D.C.: University Press of America.

Brown, James S. 1900. *Life of a pioneer*. Salt Lake City: Cannon.

Butler, Jeffrey, et al. 1977. *The black homelands of South Africa*. Berkeley: University of California Press.

Carrere D'Encausse, Helene. 1981. *Decline of an empire: The Soviet Socialist Republics in revolt*. New York: Harper and Row.

Chambers, Robert W. 1987. Poverty, environment and the World Bank: The opportunity for a new professionalism. Unpublished paper prepared for the Strategic Planning and Review Department, World Bank.

Choldin, Harvey M. 1979. Review of John W. Bennett, *The ecological transition: Cultural anthropology and human adaptation*. *American Journal of Sociology* 84:1280–82.

Clark, Colin. 1967. *Population growth and land use*. London: Macmillan.

Davies, Robert, et al. 1984. *The struggle for South Africa: A reference guide to movements, organizations and institutions*. London: Zed Books.

Debo, Angie. 1970. *A history of the Indians of the United States*. Norman: University of Oklahoma Press.

DeMallie, Raymond J., ed. 1984. *The sixth grandfather: Black Elk's teachings given to John G. Neihardt*. Lincoln: University of Nebraska Press.

Ellison, Rosemary. 1980–83. The agriculture of Mesopotamia c. 3000–600 B.C. *Tools and Tillage* 4:173–84.

Falloux, Francis. 1988. Land management, titling, and tenancy. In *Sustainability issues in agricultural development*, 190–208. Washington, D.C.: World Bank.

Fieldhouse, D. K. 1973. *Economics and empire, 1830–1914*. Ithaca, N.Y.: Cornell University Press.

Fletcher, Alice, and Francis La Flesche. 1972. *The Omaha tribe*. 2 vols. Lincoln: University of Nebraska Press.

Fortmann, Louise. 1986. Rural social organization in a semiarid African country: The case of Botswana. *Great Plains Quarterly* 6:190–201.

Foundation for the Study of Plural Societies. 1975. *Case studies on human rights and fundamental freedoms (a world survey)*. 2 vols. The Hague: Martinus Nijhof.

Goldsmith, Edward, et al. 1972. *Blueprint for survival*. New York: Houghton Mifflin.

Gorse, Jean Eugene, and David R. Steeds. 1987. *Desertification in the Sahelian and Sudanian zones of West Africa*. Washington, D.C.: World Bank.

Gras, Norman S. B. 1946. *A history of agriculture*. New York: F. S. Crofts.

Grigg, David. 1980. *Population growth and agrarian change: An historical perspective*. Cambridge: Cambridge University Press.

Hames, Raymond. 1987. Game conservation or efficient hunting. In *The question of the commons*, ed. Bonnie M. McCay and James M. Acheson. Tucson: University of Arizona Press.

Hardesty, Donald L. 1977. Nature, cultural motivations and the ecological transition. *Reviews in Anthropology* 4:120–25.

Hardin, Garrett, and John Baden. 1977. *Managing the commons*. San Francisco: W. H. Freeman.

Haring, C. H. 1947. *The Spanish empire in America*. New York: Oxford University Press.

Harris, David R., ed. 1980. *Human ecology in savanna environments*. New York: Academic Press.

Headrick, Daniel R. 1981. *The tools of empire: Technology and European imperialism in the nineteenth century*. Oxford: Oxford University Press.

Heathcote, R. L. 1986. Drought mitigation in Australia: Reducing the losses but not removing the hazard. *Great Plains Quarterly* 6:225–37.

Hills, Edwin S. 1986. History of the world's arid lands. In *Arid lands in Australia*. Canberra: Australian National University Press.

Hitchcock, Robert. 1978. *Kalahari cattle posts: A regional study of hunter-gatherers, pastoralists, and agriculturalists in the western sandveld region, Central district Botswana*. Gaborone, Botswana: Government Printer.

———. 1980. Tradition, social justice, and land reform in central Botswana. *Journal of African Law* 24(1): 1–34.

———. 1985a. Water, land and livestock: The evolution of tenure and administration patterns in the grazing areas of Botswana. In *Politics and rural development in southern Africa: The evolution of modern Botswana*, ed. L. A. Picard, 96–133. Lincoln: University of Nebraska Press.

———. 1985b. The plight of indigenous peoples. *Social Education* 49:457–62.

———. 1987. Hunters and herding: Local level livestock development among

Kalahari San. *Cultural Survival Quarterly* II(I): 27–30.

———. 1988. Decentralization and development among the Ju/Wasi, Namibia. *Cultural Survival Quarterly* 12:31–33.

Homans, George Caspar. 1960. *English villagers of the thirteenth century*. New York: Russell and Russell.

Hurt, R. Douglas. 1987. *Indian agriculture in America*. Lawrence: University Press of Kansas.

Jackson, Wes, et al. 1980. *New roots for agriculture*. San Francisco: Friends of the Earth.

———. 1984. *Meeting the expectations of the Land: Essays in sustainable agriculture and stewardship*. San Francisco: North Point Press.

Jacobs, Wilber R. 1971. The fatal confrontation: Early native-white relations on the frontiers of Australia, New Guinea, and America—a comparative study. *Pacific Historical Review* 40:283–309.

Jones, Richard Foster. 1951. *The seventeenth century: Studies in the history of English thought and literature from Bacon to Pope*. Stanford: Stanford University Press.

Jordan, Terry G. 1966. *German seed in Texas soil: Immigrant farmers in nineteenth century Texas*. Austin: University of Texas Press.

Kawashima, Yasuhide. 1969. Legal origins of the Indian reservation in colonial Massachusetts. *American Journal of Legal History* 13:42–56.

———. 1974. Indians and southern colonial states. *Indian Historian* 7:10–16.

Kerridge, Eric. 1967. *The agricultural revolution*. London: Allen and Unwin.

Khazanov, A. M. 1978. Characteristic features of nomadic communities in the Eurasian steppes. In *The nomadic alternative*, 119–26. The Hague: Mouton.

———. 1981. Myths and paradoxes of nomadism. *Archives Européenes de Sociologie* 22:141–53.

LaFarge, Oliver. 1956. *A pictorial history of the American Indian*. New York: Crown.

Lattimore, Owen. 1951. *Inner Asian frontiers of China*. Boston: Beacon.

Lewis, David, and William Reinsch. 1984. Erosion magnitude seen in Nebraska. *University of Nebraska Farm, Ranch, and Home*, special edition, 17.

Little, Peter D. 1984. Regionalism and the Great Plains. *Western History Quarterly* 15:19–38.

———. 1987. *Lands at risk in the Third World: Local level perspectives*. Boulder, Colo.: Westview Press.

Lydolph, Paul E. 1986. Comparative drought strategies: The Soviet Union. *Great Plains Quarterly* 6:246–56.

McCay, Bonnie M., and James M. Acheson, eds. 1987. *The question of the commons*. Tucson: University of Arizona Press.

MacLeod, William C. 1928. *The American Indian frontier*. New York: Knopf.

McNeill, William. 1964. *Europe's steppe frontier*. Chicago: University of Chicago Press.

———. 1983. *The great frontier: Freedom and hierarchy in modern times*. Princeton: Princeton University Press.

Madariaga, Salvador de. 1947. *The rise of the Spanish American empire*. London: Hollis and Carter.

Marks, Shula, and A. Atmore, eds. 1980. *Economy and society in pre-industrial South Africa*. London: Longman.

Marks, Shula, and R. Rathbone, eds. 1982. *Industrialization and social change in South Africa*. London: Longman.

Martin, Calvin. 1978. *Keepers of the game: Indian-animal relationships and the fur trade*. Berkeley: University of California Press.

Martin, P. S. 1967. Prehistoric overkill. In *Pleistocene extinctions: The search for a cause. Proceedings of the VII Congress of the International Association of Quaternary Research* 6:75–120.

Matley, Ian Murray. 1967. *Central Asia: A century of Russian rule*. New York: Columbia University Press.

Mingay, G. E., ed. 1975. *Arthur Young and his times*. London: Macmillan.

Mollison, William. 1984. *An introduction to permaculture*. Orange, Mass.: Yankee Permaculture.

———. 1986. *The best of permaculture*. Adelaide: Pitie.

Montagu, Ashley. 1965. *The idea of race*. Lincoln: University of Nebraska Press.

Morgan, T. J. 1892a. *Annual report of the commissioner of Indian affairs*. Washington, D.C.: Government Printing Office.

———. 1892b. *Proceedings of the Lake Mohonk conference*.

Nabhan, Gary. 1985. *Gathering the desert*. Tucson: University of Arizona Press.

Nations, James D., and Richard B. Nigh. 1980. The evolutionary potential of Lacandon Maya sustained-yield tropical forest agriculture. *Journal of Anthropological Research* 36:1–30.

Neihardt, John G., ed. 1961. *Black Elk speaks*. Lincoln: University of Nebraska Press.

Olson, Paul A. 1979. *The book of the Omaha*. Lincoln: Nebraska Curriculum Development Center.

Organization for Economic Co-operation and Development (OECD). 1985. *The state of the environment: 1985*. Paris: OECD.

Orwin, C. S. and C. S. Urwin. 1967. *The open fields*. Oxford: Clarendon Press.

Pike, Kenneth L. 1954. *Language in relation to a unified theory of the structure of human behavior*. Glendale, Calif.: Summer Institute of Linguistics.

Pike, Kenneth L., and Ruth M. Brend. 1976. *Tagmemics*. 2 vols. The Hague: Mouton.

Pipes, Richard. 1964. *The formation of the Soviet Union: Communism and nationalism, 1917–23*. Cambridge: Harvard University Press.

Popper, Deborah Epstein, and Frank J. Popper. 1987. The Great Plains: From dust to dust. *Planning*, December, 12–18.

Powell, J. M. 1986. Abideth forever? Global use of semi-arid lands in the interwar years. *Great Plains Quarterly* 6:151–70.

Rennie, Ysabel F. 1945. *The Argentine Republic*. Westport: Greenwood Press.

Rosenberg, Norman J. 1986. Adaptations to adversity: Agriculture, climate and the Great Plains of North America. *Great Plains Quarterly* 6:202–17.

Ross, Eric Barry. 1978. Food taboos, diet, and hunting strategy: The adaptation to animals in Amazon cultural ecology. *Current Anthropology* 19:1–36.

Slater, G. 1913. *The land: The report of the Land Enquiry Committee*. London: Hodder and Stoughton.

Spicer, Edward H. 1961. *Perspectives in American Indian culture change*. Chicago: University of Chicago Press.

———. 1969. Politica gubernamental e integración indigenista en Mexico y Estados Unidos. *Anuario Indigenista* 29:49–64.

Strange, Marty. 1988. *Family farming: A new economic vision*. Lincoln: University of Nebraska Press.

Strehlow, T. G. H. 1965. Culture, social structure, and environment in Aboriginal central Australia. In *Aboriginal man in Australia*, 121–45. Sydney: Angus and Robertson.

Trapido, S. 1978. Landlord and tenant in a colonial economy: The Transvaal, 1880–1910. *Journal of Southern African Studies* 5:26–58.

USDA (U.S. Department of Agriculture). 1980. *Appraisal 1980*: Soil and Water Conservation Act. Washington, D.C.: Government Printing Office.

Vickers, William T. 1988. The game depletion hypothesis of Amazonia adaptation and ten year data from a native community. *Science* 239:1521–22.

Warren, Andrew. 1985. The role of science in the management of drylands. Paper delivered at the 1985 conference of the Center for Great Plains Studies.

Watson, Steven. 1960. *The reign of George III: 1760–1815*. Oxford: Clarendon Press.

Webb, Walter Prescott. 1931. *The Great Plains*. New York: Ginn.

Wedel, W. R. 1983. Native subsistence adaptations in the Great Plains. In *Man and changing environments in the Great Plains. Transactions of the Nebraska Academy of Sciences* 11 (special issue).

White, Lynn. 1967. The historical roots of our ecologic crisis. *Science* 155:1203–7.

White, Richard. 1983. *The roots of dependency: Subsistence, environment and social change among the Choctaws, Pawnees and Navajos*. Lincoln: University of Nebraska Press.

Wilson, Gilbert L. 1987. *Buffalo Bird Woman's garden: Agriculture of the Hidatsa Indians*. St. Paul: Minnesota Historical Society.

World Commission on Environment and Development (WCED). 1987. *Our common future*. Oxford: Oxford University Press.

Worster, Donald. 1979. *Dust Bowl: The southern plains in the* 1930s. New York: Oxford University Press.

———. 1984. Thinking like a river. In *Meeting the expectations of the land: Essays in sustainable agriculture and stewardship*. San Francisco: North Point Press.

Wright, Ione S., et al. 1978. *Historical dictionary of Argentina*. Metuchen, N.J.: Scarecrow Press.

Yellen, James, and Richard B. Lee. 1976. The Dobe/Du/da environment. In *The Kalahari hunter-gatherers*, 27–46. Cambridge: Harvard University Press.

Part One

The History of
Institutional Transformation

The essays that follow, by John Bennett and Anatoly Khazanov, trace the relationship between what Bennett calls "low-energy" and "high-energy" societies in semiarid regions. Bennett begins with Great Plains indigenous people and extends his analysis to include first-generation open-range cattlemen and settlers as well a similar people on other continents. The primary contrast comes between peoples dependent mainly on local resources and those dependent on external industrial and economic systems, which may impose on the region adaptations different from those apparently demanded by nature. In the companion essay Khazanov, centering his analysis on Soviet Central Asia, makes a distinction between pastoral nomadic and industrial societies similar to Bennett's low-energy/high-energy contrast. But looking at the relationship between the mobile peoples of Central Asia and adjacent empires such as the Chinese, he emphasizes the dependence of Asian mobile peoples on sedentarized high cultures even before industrialization. However, that dependence included more equality than does the pastoralists' modern relationship to empire, because earlier pastoralists often had the edge in the military realm. Both Bennett and Khazanov emphasize the appropriateness of mobile resource-seeking regimens and other institutions to the grasslands regions—their skill in dealing with risk, flexibility of response, capacity to deal with stress, limitation of expectations, and homeostatic strength. Their systems for regulating property, governance, and resource-gathering made this strength possible. Both authors emphasize that such institutions are now virtually extinct throughout the world and describe the waste that has come with their demise. As substitutes for what has been lost, Khazanov proposes extensive pastoralism and self-determination for indigenous peoples, whereas Bennett proposes learning to live with limits and with the prospect of fluctuating population size as a sort of modern counterpart to nomadism.

Chapter One

Human Adaptations to the
North American Great Plains
and Similar Environments

John W. Bennett

John Wesley Powell's famous "Report on the Arid Lands of the United States" classified as arid lands the entire Great Plains west of the 100th meridian, and not merely the drier western half. Powell, writing in the 1870s as the Native American populations of the West were meeting defeat, had relatively modest expectations for American civilization on the Plains and accordingly emphasized the limiting effects of shortages of water and other basic resources. He acknowledged that human expansion would occur and that an economy could emerge, but it would do so only through great effort (Powell 1962). The Canadian analogue to Powell's report, written by John Palliser (1859), drew even more pessimistic conclusions with respect to the Canadian section of the Plains.

But these qualified judgments were only one chapter in a long story. Throughout the history of exploration and colonization of the Plains by Euro-Americans, the image of its environmental potential fluctuated from desert to garden, and then back again. The concept of the Great American Desert preceded Powell and Palliser, but both of their reports were followed by more optimistic assessments, as settlement proceeded through political necessity and entrepreneurial drive. By 1900 the Great Plains and the "desert" were believed to be a potential garden, with water and rich soils available in abundance, providing they could be properly developed. Then came the Dust Bowl, and the pendulum swung back to pessimism, leading Henry

Nash Smith to observe that the "West should submit to a rational and scientific revision of its central myth"—i.e., of potential abundance (Smith 1950; Blouet and Luebke 1979)—that is, back to Powell and Palliser.[1]

But the myth of abundance never died, and with intensive exploitation of plains soils and water resources, the promises would, at least at intervals, seem to be fulfilled—albeit with external subsidies as a helping hand. In larger terms, these attitudes have formed the basic realization of the process of Western settlement: *the illusion that constraints were illusory*, because North American cultural aspirations could not tolerate a sense of limits. Even Walter Webb (1932), who certainly had a sense of the way eastern adaptive styles had to change to meet western resource realities, was not pessimistic or especially prudent—all the easterners had to do was to "adapt," and things would work out.[2] Adapt, yes, but at what level of exploitation and at what environmental and social costs?

The adaptive strategies used in arid and semiarid regions by preindustrial indigenous societies are radically different from contemporary industrial strategies. The constrained attitudes and institutional arrangements such environments require if the cost of living in them is not to be prohibitive are not those generally pursued by contemporary societies, which appear to demand much greater yield from dryland resources. In analyzing this tendency toward rising expectations, I will first examine the demands that semiarid environments place on human beings, the differing adaptive assumptions of their cultures, and the implications of modern concepts of economic progress and entrepreneurial exploitation. I will turn then to accounts of American Indian and Euro-American occupations of arid and semiarid North America. The basic view is that, whereas in preindustrial cultures human adaptations are made primarily to a local physical environment and to an indigenous, self-regulating social environment, under industrial regimes adaptation requires coping with external institutional factors outside the control of local resource users. They respond not so much to nature as to culture. However, when a sustainable way of life was created by indigenous peoples in arid lands, it characteristically required a lower level of expectations than those in industrial, consumer-centered cultures. Indigenous peoples displayed a willingness to live with ambiguity and modest consumption levels, and they manifested a determination to protect their locality from outside exploitation.

ENVIRONMENT AND ADAPTATION

The Concept of Adaptation

All scientific and scholarly fields concerned with living things use the term "adaptation," each defining it in accordance with its special interests. Whereas the biological sciences interest themselves both in the genetic changes associated with natural selection and other processes in evolution, and in the physiological changes occurring in the lifetime of individuals in response to particular environmental pressures, the social sciences use "adaptation" in the sense of behavioral changes and adjustments (my main concern in this chapter). In general, when we use "adaptation" in the social-behavioral sense, we refer to the changes in the posture and activities of human beings vis-à-vis the physical and cultural environments that enable them to cope with daily life or to improve their life chances.[3] The root meaning therefore refers to defensive strategies and life-enhancing accomplishment.

Over long historical periods, adaptive behavior on the part of human populations may come to have biological significance, as in the case of tool use by early pre- and protohominids, which may have become a medium of selection for more intelligent populations. Thus social-behavioral adaptation can be either passive or active: the individual or population may simply conform to environmental constraints without attempting to alter their nature. But the active mode involves alteration, and in the course of human history this active mode has become increasingly important as a result of technological development. However, both modes seem to imply some kind of *change* in behavior: in the passive instance, the individual copes by making his own behavior conform to environmental characteristics; in the active, he changes both his own behavior and the environment itself. *Change* is therefore the most general meaning of adaptation.

Biological Adaptations

Genetic or inheritable somatic and physiological adaptations to any aspect of the physical environment are extremely hard to find in *Homo sapiens*, for two main reasons: First, human beings extensively use their cultural apparatus—technology, purposive movement, strategic manipulation—to adapt rather than changing genetically. Second (and this is really a corollary), human populations migrate, and therefore few are really long-term indigenous inhabitants of particular and specialized environments (I will mention some exceptions later). Most substantial populations of the North American Great Plains—both the Plains Indians and the later Euro-American and Euro-Canadian settlers—were latecomers, living in the region (or moving in and out of it) for only two centuries for Indians and even less for the European-derived settlers.

Concerning physiological adaptation, we are dealing principally with three processes: adjustments made by an individual's body processes over the course of his growth and maturation in a particular environment, generally called "acclimatization"; learning or devising technical, social, or ideational methods of coping; and genetic change, which can be ruled out as a significant source of physical adaptation, especially for peoples who have only recently entered specialized habitats. Both the first and second processes require two or three human generations—one needs to grow up in a particular specialized environment to become fully acclimated physiologically and fully educated culturally and behaviorally.

Although the geographic focus in this chapter is the semiarid grassland section of the North American continent, we will also look at human adaptations to environments that have moisture deficiencies wherever they may be found.[4] Since the literature on acclimatization distinguishes between hot, dry and cold, dry conditions, let us begin with the former: animals other than human beings have found ways, through evolutionary change, to cope with physical stress resulting from dry heat. Some promote evaporation through perspiration, which cools the organism—in which case a fairly reliable external water supply is needed; some insulate themselves with coats of hair or hide in the cool shade during the day; some acquire their water from eating plants, and others are never far from surface water. The problem of handling excessive heat and dryness is common to all, but there is no single method of coping. One conclusion, generally agreed upon, is that environments featuring great dry heat impose considerable stress on animals—more so than the tropical forest (Newman 1970, 12–27, esp. 15).[5] The heat is transferred from the environment to the organism in the presence of open skies, drying winds, and glaring sun; the organism has to avoid or get rid of it, and water is crucial in doing so.

Where does the human animal stand in relation to other animals? At first sight *Homo sapiens* appears to be one of the creatures least likely to accommodate to hot, dry climates since we are rather large organisms and so absorb a lot of heat. Human beings also have little body hair, so they cannot use it as insulation; they perspire freely—more freely than any other mammal, so far as we know, and while this helps them cool off through evaporation, it also requires a good water supply to replenish the loss.[6] Their skin is susceptible to ultraviolet radiation; yet the search for water and food in the sparse environments so often associated with aridity requires movement in the open country. There are a series of trade-offs here: dangerous exposure to the sun combined with movement assists in evaporative cooling; large body size means much heat, but it also promotes cooling, since the surface area per unit of body volume is minimized and therefore distributes the heat throughout a greater body mass, protecting particularly vulnerable organs.

But the biggest trade-offs of all concern the behavioral capacity of human beings to devise ingenious ways to cope with heat and drought that have little or nothing to do with basic physiological attributes: the technological and behavioral modifications mentioned previously. If water holes are far apart, you carry water in containers: if the sun is too hot, you invent cloth or skin coverings; if food is scarce, you make use of the hot, dry air or salt deposits to preserve it; and if things get too strenuous, you pack up and go to the cool, moist woods or mountains. The very incompleteness or unevenness of the human physical adaptation to aridity assists in promoting innovation and movement. Nevertheless, the human presence in arid and semiarid lands has always had problems: although humankind can exist biologically in these lands, it is at the cost of a specialized adaptive behavioral and institutional profile. The constraints are severe, the possibilities for transcending them are limited, and thus the ultimate expansion of human potential has historically been less for arid lands than for the temperate and humid lands. To expand this potential beyond the level that an indigenous population can supply from its own resources is difficult, and subsidies are usually required. In indigenous cultures, nature is the primary source of adaptive pressure, but in industrial, subsidy-based cultures the primary pressures are likely to come from the donors of the subsidies.

To return to nonindustrial adaptation for a moment, I should say a word about possible physical changes in long-term indigenous populations coping with heat and aridity. There are several classic cases of very long-term residence: the Australian Aborigines of the central desert and the Bushmen of South Africa are probably the best known. Some recent researchers with the Australian tribal groups have suggested that these populations—present in Australia for at least twenty thousand years—have a few physiological specializations regarding metabolic rate, blood flow, and perspiration (Kirk 1981, chap. 8). For example, studies seem to show that Aborigines can drink large volumes of water without distress; they excrete water more rapidly through the kidneys; they perspire more freely but can also suppress this high rate of sweating in humid environments. Hence as a rule they can travel long distances with less water than can Caucasians. There are other differences, but it is not known whether these specializations result from maturational habituation or are actual heritable traits due to selective adaptation. Caucasians reared in temperate, humid environments can readily adapt to hot, dry environments, given time and enough sense to take advice on how to beat the heat.

Concerning cold stress, anthropologists have done some research on special adaptations to cold in tribal populations like the Eskimos and the "original peoples" of Tierra del Fuego, but the results, as with adjustment to heat, are not conclusive about genetic transmission. In any case, for Plains inhabitants, clothing, shelter, and heat from fire, stoves, and such, were the main protective devices. Both summer heat and winter cold caused suffering for Euro-American-Canadians in the first year or two of settlement, according to the reminiscence literature, and deaths from exposure, particularly among the infants and the aged, were significant. But even here the reactions were due as much to economic privation and the inability to build adequate shelters or find enough food as to the weather itself.

An analysis of firsthand accounts of European farmer immigrants to the Great Plains that I made recently suggests that though some never accepted the discomforts and departed after a few months or years, acclimatization in the behavioral sense—learning protective techniques as well as accepting the situation—took at least one full year in the best of cases and as long as a generation in others. The basic physiological requirements for water, heat retention and loss, and so on, were hard enough to meet, but the sweating, blowing dust, frequent failure of garden crops, uncertain diet, and winter cold and blizzards were simply too much. The impact on human survival and health in immigrant populations was severe: infant and old-age mortality was very high; in some districts we have examined, families anticipated the death of at least one out of every three infants born there or brought into the area. The first two winters and summers were the hardest to bear, since not only was the physical stress high, but the failure of economic adaptation caused severe privation.[7] I see no reason to believe that earlier groups adapted more quickly.

The difficulty of agrarian adjustment to such lands is not new, particularly when agriculture and nuclear settlement are involved. Although arid lands with rivers running through them (Mesopotamia, Egypt, the Punjab, and some other marginal cases) were the sites of the earliest urban trading-based civilizations, most of these cultures eventually had severe difficulty with aridity. Irrigation was essential to support large populations in such lands, and most of the ancient irrigation systems eventually succumbed to siltation, salinization, or crumbling banks. The wealth that might be extracted from these lands was limited: trade was essential, and the superstructure was vulnerable to collapse when the farmers suffered prolonged drought or the deterioration of water supplies or soil fertility (Bennet 1974, chap. 2, esp. 42). These civilizations accumulated and concentrated wealth by extracting fertility from the soil (and whatever minerals the area might possess), by manufacture and trade, by subsidies from other cultured populations (often booty acquired by military conquest), and by regular episodic out-migration of surplus population that spread the limited "take" among fewer people. The economic-ecologic picture was also marked by extreme or at least recurrent fluctuation—much of which was due to natural cycles like drought or to failures in the social or material apparatus needed to maintain a constant flow of wealth or at least survival needs for a given number of people. If this pattern sounds familiar, one need only call attention to the physical, economic, and demographic fluctuations of modern populations of the North American Great Plains. The external subsidies are larger, but vulnerability remains great in the face of unchanging physical constraints.

Contrasting Adaptations in Low- and High-Energy Societies

If agricultural societies have often run into trouble in dry regions, what of tribal societies that practiced limited agriculture or none? Although high civilizations residing in arid/semiarid lands had difficulty sustaining a high level of economic output or wealth accumulation, the tribal societies, particularly the transient or nomadic ones relying more on animal husbandry than on plant crops, or those that subsisted, where it was possible, on hunting and gathering, were probably the most long-lasting and stable human cultures in these constrained environments. All of these were small, fertility-limiting populations, existing at low levels of economic output and consumption. The migratory adaptation and communal land tenure and use fitted the specialized and sparse resources, and their habit of movement and dispersal allowed easily damaged resources to regenerate.

Such societies can be considered "low-energy" adaptations. "Energy" here refers to the amount of effort required to extract a product, whether nutritional, material, or social.[8] The word also can refer to the secondary energy-producing potential of the product itself, after extraction and processing. Low-energy societies use human and animal muscle power, and with this limited technology they can extract just enough energy from the habitat to provide sustenance and shelter for the resident population, with little left over for luxuries. Such societies were relatively self-sufficient, although a certain amount of trade and barter always occurred between groups. They were generally "in balance" with the environment, which means that their technology permitted a regeneration of resources. However, there was always a degree of cyclical use, overuse, and recovery, and in some instances, as with the Plains Indians' habit of driving bison herds by firing grass and thereby further reducing tree cover, there was significant long-lasting impact. On the whole, low-energy societies were ideally suited to the highly variable and constrained physical environments of arid/semiarid lands.

In a "high-energy society" the product obtained from the resources is produced with technology supplying a larger quotient of energy than that available from human or animal muscles. Moreover, the machinery, generated power, or chemical transformations permit a riskier use of resources, particularly in situations where regeneration is difficult or impossible. Characteristically, such technologies of production supply larger populations than those immediately engaged in production. Markets rather than sustenance are the main objective, and typically the local population loses control of its resource base, or at any rate the people no longer produce for their own subsistence.

Thus high-energy uses of arid/semiarid lands have a proclivity toward resource abuse, primarily because their extractive processes are extensive and exhaustive and the external demands for such resources are so high. The local population must survive on the cash benefits their resources can bring from outside buyers, who seek a supply with little or no concern for sustaining the yield of the resource. In agriculture, this tendency to push such specialized and often marginal resources beyond regenerative capacity is encountered in grasslands in the form of overgrazing; in cropping areas, in the tendency to pump groundwater excessively in order to irrigate; in the localized impoundment of streams that formerly watered extensive dry areas; and in the disturbance of friable soils in agriculture and urban fringes, causing wind and water erosion. When multiple causes and effects result in extensive loss of vegetation, surface water, and groundwater, the term "desertification" has been applied (see later discussion). Although the term lacks scientific precision, the best-known case is the Sahel, the band of scrub vegetation constituting the transition between the true Sahara desert on the north and the forest on the south that has undeniably undergone increasing desiccation, much of it due to overgrazing, the opening of marginal lands to cultivation, and an unfortunate drought cycle. The "advance" of desert conditions southward into the Sahel is variously estimated, depending on the year and the district, but there seems little doubt that human activities, stimulated by development projects, increasing concentration of population, and more vigorous technological interventions, have accentuated land and plant resource degradation.

Variability of Semiarid-Variable Resources and the Problem of Adaptation in Indigenous and Industrial Cultures

Perhaps the most crucial aspect of arid/semiarid environments is their highly variable climate, water, and soil (although the nature of this variability differs by location). Thus, semiarid grasslands in the Northern Hemisphere are subject to a great seasonal temperature range and highly variable rainfall, both characteristics responsible for human physical stress and also high-risk agriculture. Desert regions in the tropical latitudes have considerable night/day temperature extremes; indeed, tribal nomads in these environments require not only acclimatization to dry heat, but also tolerance of very cold nighttime temperatures (the voluminous robes of the Bedawi protect against both heat and cold). Most arid/semiarid lands experience drought as in some sense a permanent condition, but really severe droughts appear at intervals not subject to reliable forecast: they may last a year or two, or they may endure for a decade or more, as they did in the Great Plains in the 1930s. This variability creates special problems for high-energy adaptations to these lands, since production for markets demands responsive flexibility in changing the nature and the volume of crops to suit prices and costs—changes that are always subject to environmental constraints. Variability simply makes the response less confident.

Since variability is as much a characteristic of these environments as drought, we need the special acronym: arid-semiarid-variable, or ASAV. For the Great Plains of North America, this implies that humans must cope with heat, cold, drought, torrential rains, blizzards, and the highly variable temporal manifestations of them all. In addition, the soils of these regions are extremely varied owing to the savage action of glaciation and glacial melting, which created vast lakes and rivers, moving the particles of soil according to current speed and topography and segregating them into sand, gravel, silt, and other forms with differing agricultural potential. This was of no great significance for pastoral nomads like the Plains Indians, but the homesteading farmers found it hard to cope with.

Culture is always part of the adaptive process, since the concept of culture may be defined as the proclivity of humans to create new and changing ways of dealing with nature and each other. Even more important, it is the human ability to remember these ways and store them up for future use. This accumulation of precedents and styles and their changes permits human beings to anticipate future needs and to call forth both innovations and traditions to suit particular demands or to handle emergent constraints. Thus humans are not in any simple way "determined" by physical phenomena but cope with or adapt to them. However, this dynamic quality of human adaptive behavior does not obviate the need to restrain ambition or cultural interests when environments create difficulties or when culture does not permit adequate management. So in a qualified sense the environment does set limits, but it also provides opportunities—it is often difficult to know which is which, since limiting factors often can stimulate innovation that transforms them into opportunities. This is as true for low- as for high-energy societies.

For example, Amerind populations across the North American continent were markedly different in cultural style, and ethnologists have long recognized a role for the environment in helping bring about such differences in low-energy tribal cultures (Kroeber 1953).⁹ The Plains Indians, in adapting to open spaces and to horse and bison pastoralism, evolved a vigorous, mobile, aggressive style, a common type for ASAV lands the world over. This style can be compared with its virtual opposite: that developed by Indians in the lush environments of New England. William Cronon attributed the rather leisurely lifeways of these tribes to the ecological abundance and diversity of the environment, both within and between the markedly different four seasons (Cronon 1983). The Plains Indians, on the other hand, had to cope with less diverse and more specialized resources and with an essentially two-season annual round: short, hot, dry summers, and long, cold winters.

In the Great Plains, where physical resources were widely dispersed and one, the bison, was mobile, the Indians were required to move long distances to obtain the raw materials and climates they needed for survival or for different types of encampments suited to different seasons (open-country sites during the hunt; sheltered permanent camps during the winter, etc.). The New England agricultural tribes, able to obtain locally most of what they needed at all times of the year, could build permanent villages and remain in them indefinitely, although some groups moved their villages at long intervals to permit soil regeneration or simply to find new opportunities.

This "eastern leisurely style" also extended in some degree to the European immigrants. When the Europeans came, the lush eastern lands permitted them to reproduce the institutions and settlement patterns familiar to them from European homelands, since the environments were so similar. However, the Euro-American settlers who came into the Great Plains of Canada and the United States were confronted with an unfamiliar milieu and had to innovate—a theme explored by Walter Webb (1932) and Frederick J. Turner (1958)—and also oversimplified by both.¹⁰ Nevertheless, in the early days of western North American settlement, attempts were made to reproduce eastern institutions, such as the rectilinear land survey system that ignores the complex drainage patterns and sparse water supplies of the West.

The Euro-American-Canadian settlers in the Northeast could establish permanent communities immediately, since relatively little energy had to be expended locating water, good soils, game, and fuel. On the other hand, in the Plains the search for these basic necessities of settled life dragged on through two or three generations of settlers. In my work on migration to and within the Northern Plains, I have found that fewer than 20 percent of all settlers in the more northerly portions stayed put—that is, found a homestead on the first try and stayed there. Most kept wandering, looking for more salubrious climate, soil, water, and neighbors. For many the final settlement locus was a last resort—weary of the constant movement, they came to realize that there was really nothing better. Resources in many plains areas never seemed to come all together: if the well was good, the soil might be bad or the topography rough.

But such a limited region is not likely to be greeted with limited expectations in an industrial society. In an entrepreneurial economy, fueled by high aspirations and the notion of continual progress, adaptation comes to be viewed as a directional change process; by "adapting" things get better; by solving problems life becomes easier, and thus progress is served as well as validated. However, if resources are only marginally sufficient to realize these goals, subsidies need to be found, and if subsidies are not forthcoming, then the resources must be *made* to yield, at whatever cost.

Once this course of action is set in motion, the process may appear to change from a directional one, with ever-rising vectors, to a cyclical one, with ups and downs, boom and bust. And the question whether these cyclical movements are simply squiggles on the generally ascending curve of prosperity, or whatever, is difficult to answer, particularly when some of the changes, like resource degradation, appear to be accelerating downward and creating costs. As prairie soils lose fertility in continuous cropping, expensive fertilizer must replace it, and the cost of replacement may be met by rising economic values elsewhere, such as increased federal subsidies or higher food prices. But all of these factors are hard to define and difficult to sustain. Consequently cyclical movement, and uncertainty as to the precise phasing of the cycles, becomes the pattern.

Moods of optimism and pessimism, caution and daring, innovation and resignation accompany these movements. If one uses the modern North American Great Plains population as a model, one finds that these fluctuations are geared not only to climate and economic conditions, but also to the human generations. The first generation of settlers was basically optimistic,

ographers. Working with the available local resources to obtain food and raw materials, these are mainly go-it-alone societies, and hence all the more admirable from the standpoint of modern adaptations to ASAV lands, which depend heavily on external energy inputs. (3) Doing it on your own also means exercising choice over which resource to specialize in, even though the resource base is limited compared with other habitats. (4) Nearly all of the tribal adaptations to these environments involve more than a single microhabitat; these people have worked out complex transhumant adjustments to a variety of differing seasonal, topographic, vegetational, and environmental features in the general region: the human "possibilist" adaptations are analogous to the ecological principle of diversity.

We can illustrate this with a brief summary of the available habitats in North American ASAV environments: desert oases; mountain meadows; shortgrass plains traversed by intermittent sparse streams; midgrass prairie with frequent streams with gallery forest and shrublands; major river valleys with associated forests, bottomlands, and tributaries; drylands covered with shrub growth; drylands with sparse and xerophytic vegetation ("deserts"); and upland regions with parkland vegetation (Shelford 1963). The list is considerable, and each habitat harbored distinctive forms of human existence; foot hunters of wild game; horse-raising bison hunters; agricultural pueblos on mesas; streamside agricultural villagers; hunters and horticulturalists in bordering forests; and many combinations or blendings of these types. Some groups like the Crees might live in the prairie areas but return to the forests for hunting seasons. Bison-hunting horse tribes like the Crows might have agricultural relatives along river valleys, with frequent visiting and interaction between the sections. Many tribes had dual or even triple adaptations, accustomed to a variety of subsistence undertakings and to using a large variety of animal and plant substances from a variety of habitats.

Changes occurred constantly. The relatively open topography and sparse vegetation of many regions permitted a great deal of movement and contact: war, raiding, trading, political negotiation, constant travel, and exploration of new foraging grounds were typical of plains and prairie and mountain foothill people.[12] Learning new ways took place regularly, and cultural traits and concepts spread across large geographical areas, with constant change and permutation. The bison hunters with horses derived from Spanish herds—the so-called typical Plains Indians—were of quite recent origin and derived from bands and villages of tribes of varied subsistence types on the borders of the plains. There was no real shortage of food anywhere in the prairies, plains, and deserts for those who knew. The Shoshone Indians of the Great Basin, roaming in one of North America's most "barren"—from our point of view—lands, found no difficulty in supplying themselves with adequate vegetable and animal food and raw materials with their technology.[13]

Hunting-and-Gathering Adaptations

Thus the earliest indigenous adaptations to ASAV lands are called food collecting, or hunting and gathering (HG) by ethnologists, and there is no doubt that they represent the earliest forms of hominid adaptation to all physical environments.

HG peoples vary in ideational culture and social organization, as did all tribal entities, but the mobility associated with HG adaptations, and the total reliance on foods collected from the natural world, appears to have conferred certain broad similarities. The sexual division of labor, with men hunting animals and women gathering vegetable foods, is nearly universal. Most HG bands had base camps and wide-ranging foraging territories, the areas divided among the various bands by customary arrangement. Tools were all of the hand variety, but the types of projectile weapons, snares, traps, storage facilities, and even in some cases knowledge of simple irrigation to encourage the growth of wild plants constitute a sophisticated armament quite adequate to the tasks and appropriate for the support of a small population.

These peoples were, above all, oriented to large geographical spaces. Lacking nucleated permanent settlements other than the base camps, and lacking a concept of fixed land tenure, they came to know the signs and markers of special boundaries and places with great precision. Territories were traversed by many trails and tracks; the foraging individual or band was never "lost." The tracking capability of HG peoples was impressive and has been featured in modern novels and stories. The kinship social organization of HG populations was not simple but often, as in the case of the Australian Aborigines, enormously complex and included hundreds of persons dispersed over very large territories. Ethnological accounts of these peoples have emphasized the ability to locate exact positions in social networks, even when the individuals and groups had never previously met or interacted.

The mode of HG adaptation was risky in that severe drought and other interruptions of normal physical resource cycles could induce starvation or severe privation. However, the technology and modes of resource use were adequate to support optimal populations—perhaps larger numbers, in some instances, than could be supported with the undeveloped agricultural or ranching modes of production used by the earliest European-derived settlers in ASAV lands. In some regions, like central Australia, this latter situation would be true even today. However, the intervention of modern settlers' high-technology economies has greatly modified HG adaptations, and they are becoming extremely rare everywhere, perhaps effectively extinct. Settler occupancy of the land destroys the "food habitat" of HG peoples just as it destroys the natural habitats of animals and plants. HG humans are an endangered cultural species.

At the same time, the environmental movement as a worldwide ideology and action program has produced new threats to the continued existence, however qualified, of HG and other traditional adaptations. Friction between environmental campaigners and indigenous tribal peoples exists in many places—noteworthy cases are the Columbia River, with regard to fishery rights, and central Australia, with regard to hunted animals. To pursue an indigenous HG strategy in greatly restricted territory accentuates the problem of species destruction, and such techniques as brush firing reduce plant cover, both things alarming to environmentalists. But this problem is just one among many; the solution to the existence of remanent populations of indigenous HG people, as well as herds and agriculturalists, is nowhere in sight. It was and always will be a *political* issue, since it involves rights, property-productive resources, and ideology, and such considerations will tend to dominate biologically rational adaptive formulas.

Pastoralism

Through portions of Africa, the Middle East, and Central Asia there remain substantial indigenous populations of people who lack permanent nucleated settlements and who follow or drive their herding animals from pasture to pasture. This is a fundamental and long-lasting mode of low-energy adaptation to ASAV lands, derived from borderland agricultural civilizations, or in the more ancient instances, an autonomous mode of adaptation dating back to prehistoric times. Many such populations, particularly those in Central and Southwest Asia, retained contact with settled civilizations, and consequently they enjoyed cultural standards of considerable sophistication.[14] The ingenuity with which Asian pastoralists, especially, managed to translate these standards into portable form has been a favorite theme of ethnologists; the large substantial, insulated, yet easily collapsible tents, horse trappings, leather trunks, metal jewelry, and intricate, well-tailored clothing attest to the ability of these people to live a comfortable life and accumulate wealth and possessions, all in the absence of long-term permanent settlement or, what is most significant, a concept of wholly owned land or space. This transiency was shared, of course, with the pastoralists of most of Africa, but in other respects the African tribal herders had a simpler material culture, perhaps because they had weaker connections with the older civilizations. (The African pastoralists are generally considered "tribes"; the Central/Southwest Asian and Middle Eastern groups are often called "tribal nations.")

In North America, pastoralism was represented by the Plains Indians after the seventeenth century, when they acquired horses. Ethnologists once debated whether to consider these people "true pastoralists," since in contrast to the African and Asian pastoralists, they did not breed cattle. However, this was really a quibble, since the issue is the extent to which the pattern of human existence is shaped by movement with or because of herding animals. The bison had their own movements but were frequently driven or guided by the Indians for slaughter. Moreover, the Indians were active horse breeders, breakers, and riders. The Plains horse culture was relatively recent, with the animals descended from strays from Spanish herds in the Southwest. It was characterized by a strong set of distinctive technological, military, social, ritual, and aesthetic patterns that overlay the differing cultural elements of the component tribes, derived from different stocks and subsistence adaptations on the margins of the Plains. Thus nearly all tribes had an annual collective ritual, the sun dance; all of them used transportable dwellings, furniture, and clothing; they all established societies of young males for war training, the hunt, or fraternal ritual; and of course there was the technical dependence on the products of the bison.

As the Indians gave way to the Euro-American and Hispanic movements into the Plains, cattle displaced the bison, but horse-dominated pastoralism continued under new auspices. The initial range livestock phase consisted of driving herds from Spanish settlements northward to supply mounts, meat, and hides for miners and other settlers; this soon evolved into open-range horse and cattle raising but was equally rapidly replaced by fenced ranching—the latter an adaptation to the land survey and the opening of various sections of the Plains to freehold agricultural settlement. The open-range version resembled Asian pastoralism in many details (although of course the cultural base was very different). The range horse breeders and cattlemen were members of settled borderland societies who elected to take up a pastoral existence and yet retained the basic outline of urban culture. While they always had a home base, most of their life was spent on the range, supervising and driving the animals. Like the great khans of Central Asia, the horse/cattle barons might frequently entertain visiting and exploring members of financial and social elites from distant entrepôts or foreign countries. And both scorned the farmers and villagers yet were more than willing to take title to land and invest in urban pursuits when civilization moved their way.

The range cattlemen practiced rudimentary conservation by not concentrating the herds of different owners in the same grazing areas, through conservative or even quasi-military (armed cowboys, vigilantes) action. In addition, the adaptation was similar to Old World pastoralism in its intermittent, *surficial* use and exploitation of grasslands.[15]

In contrast, *fenced ranching* required management of both water and forage, since the herds had to be confined to designated areas and hay raised as a crop for winter feed. *This transformed open-range pastoralism into livestock agriculture.* That is, there was an absolute necessity to take control of the finite resource base and create a new rhythm of use and regeneration under the control of human beings (i.e., the "ecological transition"; Bennett 1976). Irrigation, tame-seeded forage plants, fenced grazing areas, appropriately located salt licks, shelter for brood cows and calves in the winter—and so on; all these were required now but were things the old range cattlemen would have regarded as silly or at least unnecessary, given the open availability of enormous areas of grassland or scrub.

The movement to fencing changed everything. *Mobile* resource utilization differs ecologically, fundamentally, in its use of nature from adaptation based on a fixed tract of land that must be made to produce year after year. ASAV lands have always been naturally suited to mobile resource utilization regimes, and attempts to make them conform to fixed and repeated production strategies have always experienced difficulties, and always will. The constraints are severe, and the ability of these environments to respond to production pressure is limited. The modern world exploits ASAV lands by requiring them to produce large volumes of a relatively small number of products, for the maintenance of indigenous populations and for populations at indefinite remove from the ASAV environment. When such external demands were weak, the pattern of adaptation was locally diverse, conservationist on the whole, and geared to subsistence. It is true, of course, that there were exceptions in the area of conservation, like the burnoff of grasslands and local overgrazing, or the hunting to extinction of certain mammals, but on the whole the human populations remained modest and the pressure on resources was sufficiently benign to permit regeneration.

The attempt to force or persuade pastoralists to settle down and become ranchers has fostered considerable research on the way physical resource use was geared to social organization and symbolic ritual. The hunt, the raising of livestock, observation of the seasons, food preparation, and other things constitute a style of "cultural ecology" much less evident in contemporary adaptations. The intricate interrelation between human activities and beliefs and the physical environment became more than an academic subject when Third World countries attempted to convert their inhabitants to commercial agrarian regimes. The failure and confusion attendant upon this effort have been particularly evident in Africa, where the attempts to convert pastoralists into limited-pasturage tenured ranchers had poor results despite the investment of several million dollars in development funds and loans.

At the root of the failures lay the inability to solve the resource-tenure problem; encouragements to increase production and improve breeds were on the whole added to the migratory system of pasture use and the customary basis of agreements between herders to respect use rights in a complex rotating sequence. It was expected that by increasing herds and improving market sales, pastoralists would automatically restrict herd size and begin to settle down into limited grazing areas and develop forage production as they did so, but this did not materialize anywhere in time to avoid devastating overgrazing of the ever more limited range—limited by government and private appropriations for alternative uses, such as game reserves, commercial wheat schemes, and the like. The problem was based on the fact that

though pastoralists used land in common, with intricate customary agreements, they owned the herds as individuals or families. That is, the more-or-
less communal land use or control system was not matched by communal
ownership or management of the grazing species. The result was that when
government encouraged increased production, this was done by individual
herd owners, but on grazing lands that were constantly diminishing. Livestock was also a form of wealth—a custom that dies hard; the system breaks
down under development pressures, and the herders are forced into the
cities to swell the numbers of the unemployed and indigent.

No better example than these attempts to "develop" pastoralism could be
found to illustrate the consequences of upsetting the careful balance between
social institutions and resource ecology among the traditional users of ASAV environments. It also underlines the precarious and exploitive nature of attempts
to use these lands with the same set of profit-making, high-production institutions devised, over the centuries, for humid environments.

Industrialized Agricultural Settlement in the Plains

Industrialized agricultural settlement of North American ASAV lands began
in most areas in the second half of the nineteenth century. It too required
special techniques.

Viewed from the comparative perspective of the humid East and Europe,
the problems of the ASAV West center on the *uncertainty* derived from the
scattered resources. Since nothing can really be counted on—streams dry
up part of the year, rain falls on your neighbor's fields but not on yours, your
field may be loamy and your neighbor's pure sand, and so on—it is necessary to devise special techniques for farming; for example, hedging, innovation, experimentation, the courage to convert a constraint into an opportunity, these were and remain essential in trying to make a living off the
country. One risk was overreliance on a particular physical feature such as
soil that was converted into a resource and that in turn eventually was degraded. Such overreliance and specialized production made settlers financially vulnerable and ultimately culminated in a boom-and-bust economy
for many localities. The attempt to convert an inherently variable resource
base into a monocultural production system also resulted in a loss of biological diversity—the extinction of many useful plants and animals.

Uncertainty, from a behavioral standpoint, means that the expectation of *particular* desired outcomes is low; that they are a matter of low or ambiguous probability. When the probability of occurrence—for example, the annual rainfall needed to produce a crop of given potential—is not known with precision, the uncertainty quotient is high. The extent to which the probability can be calculated measures the amount of risk; if the probability of good rainfall is 50 percent, one may know one's chances of obtaining a good crop are fifty-fifty. Thus the settlers in the Plains—or the Indians, chasing bison herds that fluctuated in numbers and location—had to learn to calculate risk subjectively. And when scientific knowledge of the processes involved was low, experience and the calculations derived from that had to substitute and gambling strategies were necessary ("seat of the pants" farming).

A second important concept is *flexibility*. Settlers in the East might be able to make a good living by doing the same things every year, since they could depend on resource potential, but in the Plains one had to be prepared to shift to different crops or entirely different strategies. If cropping became too expensive because of the need for fertilizer and such, one had to shift to livestock or look for oil on one's property. Experience counts: the longer the producer copes with the situation, the more *alternatives* he may learn about and experiment with. Indian grassland hunters had to be able to shift from large to small game during heavy winters or severe drought; grain farmers have to disperse their holdings across a wider region to maximize the chance of obtaining good rainfall or better soils.

Flexibility in the face of high uncertainty also means the need to accept *trade-offs* between alternatives, when the possibilities of realizing one goal may be compromised by the need to fulfill another. In an environment of chronic marginality and shortage, people learn to do what they can with what they have and to juggle competing needs. This also means that *opportunity costs* are generally high: the cost of shifting from one alternative to another may be higher than one can or wants to bear. However, the necessity is there—one changes or one does not survive. Thus the risk factor again: beg, borrow, or steal to change to the new alternative, and find some way to pay the costs. In some circumstances, however, it may be wiser to "sit it out" rather than attempt the shift to a new course of action, particularly when the alternative is especially costly in either financial or social terms. Constant movement from one homestead to another—so common among the early settlers—required weighing the cost of staying or leaving and the probability of finding an even worse site in the next county or state. Still, the chance might have to be taken if the current site was not productive and the family's welfare was at stake. Essentially only three alternatives were available: stay, try a new place, or go back to the original homeland (or some intermediate location).

Accumulation was another strategy that had to be exercised. Cash was essential for settlers—in the early days, its only source might be wage labor, and pressing as most needs were, full satisfaction often had to be deferred in order to acquire a small reserve for future needs, including the expenses of a possible move, the purchase of more land, or a necessary machine or horse. Similarly, Indians had to exercise these prudent strategies—not only on the Plains, but almost everywhere. The preparation of dried, portable high-protein and vegetable foods and the digging of storage pits in encampments are familiar to ethnologists and archaeologists. The heavy reliance on "futures"— that is, the annual return of the bison herds, the flowing of spring floods, or the flowering and fruiting of plants—meant that some preservation, storage, and transport of surpluses was essential in case of a cyclical failure.

None of these behavioral survival strategies were or are unique to ASAV lands; they are the stuff of everyday practical life and subsistence adaptations everywhere and in all climates. However, the ASAV situation enforced them with a vengeance; the marginality and unpredictability of resources meant that the residents of such lands, whoever they might be, have had to be both prudent and daring in turn, depending on the fluctuating resources.

COMMUNITY PATTERNS AND PROBLEMS IN ARID/SEMIARID REGIONS

Whatever the settlement pattern, some community problems persist across changes in methods of adaptation. For example, I noted previously that the fate of communities is bound up with the problem of small and dispersed settlement. In recent years this has been summed up in the concept of *sparse-lands*,[16] the term referring mainly to regions settled by emigrants from the more densely populated areas of western North America, Asiatic Russia, Australia, New Zealand, Argentina, Brazil, and northern Scandinavia.

Though this list contains a wide variety of habitats—temperate or tropical lands, rain-forest tropics, and the subarctic—the essential factor is the availability and convertibility of resources; energy levels are not sufficient to permit dense settlement and the financial cushion that uncertainty requires.

But social phenomena, not resources, are the topic of this section. How do people get along with few and distant neighbors? What substitutes for the more intimate social contacts and networks or high-density population? How do people adjust to the social deficits? If they do not find ways of compensating locally they seek help, and hence the practical problem of sparse-lands for national governments is how local systems demanding more substantial benefits can be financed: Are jobs to be provided by government investment in productive business, or are people to be encouraged to migrate to better-endowed regions? How are the indigenous relict populations, like Indian reservation communities, to be handled? Do they deserve the same benefits as the settlers?

To begin with the Indians, the nomadic aboriginal life had its own problems of association. The long periods of the hunt, when the tribes broke up into bands, created a need for wider contact—especially to seek marital partners, since bands were for the most part exogamous and composed of single kin groups. The Sun Dance was the principal adaptive response, a vast group ceremony or festival when all the bands came together for days, to enjoy one intense period of socializing when friendships were resumed, mates found, and the kin networks reaffirmed.[17] Some of the patterns were based on the social organization and cultural interests of the pre-Plains period for many groups, who had forged their culture in more settled life in the richer environments on the fringes of the Plains. As with the Euro-American and Euro-Canadian settlers, accustomed to the intimate associations of small towns and more densely populated rural neighborhoods, ways had to be found to resume this earlier social existence in the wide open spaces. Adaptation to ASAV lands does not necessarily mean one starts from scratch—it is a matter of reproducing earlier forms of social life but modifying them to cope with the changed environment.

Thus, whether we are dealing with nucleated Euro-American settlers or tribal revitalists, we find that people who came into ASAV environments with their own customs, forged in different climates, had to recreate new versions of these traditional forms in a resource context that tends to work against human association. The women who wrote the reminiscences of homestead life almost universally complained of loneliness, but they also described their joy and contentment at the occasional opportunities for socializing—the monthly dance in the schoolhouse, the sodalities and clubs, the long drives through inclement weather in democrat wagons or the Model T, the chances at regaining human contact that made it possible to endure the solitude and hardships. Some simply gave up—the isolation was too much for them, and population fluctuation in the Plains was not entirely a matter of economic difficulty. Many of the settlers left because it was impossible to reproduce the social life they desired.

Settlement and sedentarization also had a profound effect on the Indians. The shift from mobile horse/bison pastoralism and transient use of resources to settled reservation life was a traumatic episode to which the indigenous people of the Plains have simply never fully accommodated. As early as 1932, in her classic study *Changing Culture of an Indian Tribe*, Margaret Mead observed that although the Omaha reservation she studied required agriculture and other sedentary economic pursuits to provide an income, the ideas and habits of the male members of the tribe remained firmly rooted in the nomadic hunting, warring past. Village life was not for them; cultivating fields was seen as an insult to a real man, essentially women's work. The kind of associational life represented by permanent settlement seemed dull and vapid compared with the comradeship of hunting and war parties, the fraternal societies, and the annual ceremonies. The social organization and ideology of pastoralism have died hard; the adaptive shift is still not complete in many reservations. The delay of course is also partly attributable to bad policies on the part of the federal government: dishonesty and prevarication, failure to fund adequate education and agricultural training programs in the early days, confused land allocation and tenure rules, and misguided welfare procedures. But in general the difficulties underline the vast cultural gap between nomadic and settled life and also the difficulties with nucleated settlement patterns and existence for all peoples who must establish themselves in ASAV ands.

The villages and towns established by Euro-American-Canadians have not been exempt from "sparselands" problems. The expectations of settlers familiar with viable nucleated communities are not easily met in these lands. In central Australia, the United States West, and northeastern Brazil, villages and towns were founded immediately after colonization; in all three countries, maintaining these communities has been a constant struggle. In North America villages were constructed overnight along the railroads, as division points, service centers, settler ports of entry, and pickup spots for farm produce. In western Canada, villages were established about every eight miles along the Canadian Pacific Railway main line; today fewer than half these communities remain populated. Yet about half *do* remain, many of them clinging to existence with minuscule populations of aged retirees or the managers of a handful of service facilities. One can deplore the losses, but the other side of the coin is the amazing persistence of villages even when most of their economic rationale for being has withered. Nucleated settlement is hard to kill, even in ASAV environments.

The basic issue appears to be the existence of linkages—social and economic—between the nucleated settlements themselves and between the settlements and the more densely settled and prosperous regions on the margins. I have underlined the importance of relying on the external inputs if human adaptations in ASAV environments are to provide for their inhabitants at a level exceeding the productivity of the local resources; this can be done for nucleated settlements *only if they depend on and exchange with each other*. In the capitalist environment of North America, many such communities tend to reach a level of equilibrium that does not change for a very long time, even though the national economy undergoes many changes. The communities achieve a relative stability provided by connections with external economic resources. The resulting geographic pattern is analyzable with *locational theory*, as developed originally by Wilhelm Christaller and elaborated subsequently by many workers.[18] A hierarchy of communities of varying but rather precisely calibrated sizes forms a grid on the land surface, and the grid takes predicted shape best in a relatively level semiarid environment with an established agricultural economy, such as the northern plains of North America.[19]

Such a stable community is the town of Maple Creek, in southwestern Saskatchewan, a tertiary service center community in the heart of the Palliser Triangle, the driest portion of the Canadian section of the northern plains (cf. Bennett 1969). Maple Creek has had a population of about 3,000 permanent residents for the past eighty-five years, varying up and down by only a few hundred souls. The number of locally owned businesses has averaged about twenty-five for this same period. In periods of prosperity this number can increase by fifteen to twenty, with the excess disappearing within a decade or so because of dwindling financial resources. With the help of the federal and provincial governments, the town of Maple Creek has gradually accumulated adequate medical services and facilities for the aged and retirees. The basic businesses serve the country agricultural community: grain elevators, farm machinery dealers, automotive repair, hardware, and so on. In recent years several chain stores have moved in and currently are doing well. In 1969 I calculated that the annual value of all external government financial inputs into the town totaled about $5 million. By the 1980s this had at least doubled.

The town persists despite the presence, sixty and seventy miles to the west and east, of small, secondary service-center cities with abundant wholesale and retail services. The only type of businesses in Maple Creek to have become extinct in the past twenty-five years were two or three locally owned groceries, now replaced by co-op store and the IGA, both chain operations in a sense but locally managed and operated. Maple Creek's persistence is at the expense of several primary village centers located on a periphery of some ten to twenty miles in all directions; these communities—those still on the map—have declined to little more than places to pick up mail or fill a gas tank. Thus the survival of small service towns like Maple Creek is evidence of the workings of locational settlement patterning and processes in the age of the automobile.

But these declining or even defunct communities still display cyclical movement from time to time. The existence of a framework—an infrastructure of buildings, streets, business locations, churches, and so on—becomes a magnet, even in conditions of low-level needs. Small entrepreneurs, or service agencies, are inclined to risk a start in such a frame even though the risks are high and the expectations minimal. Or the regional school authority may decide to invest in a new central high school, and this in turn attracts residents and small businesses. So the community begins a modest revival. But these revivals go just so far; their duration is based on the specific needs served, not on automatic growth dynamics. Since wealth is hard to accumulate in ASAV environments, people must learn to live with specialized resources and a more modest scale of consumption. Boosterism persists—it is an inherent symbol in North American culture—but one feels increasingly that people in these utilitarian and circumscribed settlements have come to understand it is mostly a symbol, and underneath is a growing realism and comprehension of the constraints governing existence in Plains settlements and the "social deficits" these create (Kraenzel 1967).[20]

Fort Benton, Montana, is another example. Founded on Missouri River transportation, Fort Benton today is no longer the entrepôt it was in 1870, and in the 1950s it nearly died. But it remains a service center, thanks to its function as a magnet for agricultural settlers in the 1880s and the early twentieth century. Its great trading companies are long gone, but in their heyday they serviced not only river and overland trade and traffic, but the mercantile needs of the farm and ranch settlers. For some communities like Fort Benton, tourism continues to supply a small but reliable income. Fort Benton, like Maple Creek, is helped by its intermediate location between Great Falls to the south and Havre to the north.

We may conclude that in a modern economy the settlement of ASAV lands will be closely guided by functions and services, and less so by human desires. That is, will and intention must take their cue from the specialized possibilities and opportunity structure, not from visionary or ambitious energy. The successful community builder in this situation is the person who perceives a "niche" opening up, however small, and manages to obtain enough energy in the form of social action and money capital and the interest of outside suppliers to start the ball rolling. Thus the entrepreneur in the ASAV settlement process is by no means dead; he or she remains as vital as in the early days of pioneer and frontier settlement, but the task now is continuity, maintenance, and controlled revival.

Given this survival context, it is not surprising that, in the belles lettres produced on Great Plains subjects, one persistent theme has been materialism: devotion to humdrum affairs and hard work and distrust or neglect of the arts and things of the mind. Intellectuals flee these communities, then in later years write wistful or nostalgic but basically critical appraisals of the deadening effect of life in the little towns. Even Wallace Stegner, willing to invest a major portion of his career in writing novels and memoirs based on his boyhood in such a community, with its colorful characters and adventurers, nevertheless was unable to say anything positive about cultural matters (Stegner 1962, 306). Nobody should ask prairie villages to produce resident geniuses; this is not their function, and it never will be so long as our continental system of production remains what it is. One must find other values in these communities: the steadfastness of people who have lived with less and yet nourished hopes of better things; people who are willing to work hard for little return by other standards; a culture that, while emphasizing individual accomplishments, does not stint in providing help and cooperation. The values are social, not intellectual.

In some ways we now have a more positive appreciation of the old cultures of the indigenes. Fear of the terrifying Plains Indians has gradually given way to admiration of their fortitude and determination in defending their world. Belles lettres has glorified the works stimulated by Black Elk, or written by Louise Erdrich, N. Scott Momaday, James Welsch, Harold Cardinal, and others, and the very nomadism of the Indians can be romantically appreciated precisely because it is so different from our own abiding attachment to sedentary nucleated settlements. And now, of course, the Plains Indians are economically, if not wholly psychologically, committed to this system. The grand paradox of our culture is that while we are so attached to nucleation, we find difficulty appreciating the values and meaning of small-settlement existence. We tend to make fun of it, deplore it, and perennially decide it is on the way out. I think it will be here for a long time to come and will continue its pattern of rise and fall, of decline and revival. We also, of course, deplore this fluctuation as debilitating and destructive of human aspiration; yet there is another side: the courage and persistence of people who stick to something even in times of despair and make the best of it.

EPILOGUE: ADAPTATION AND MALADAPTATION

Earlier in the chapter I introduced the term *desertification* as referring to a congeries of degradational effects in ASAV environments resulting from combinations of human and natural causes.[21] Although the concept seems valid for parts of the Sahara-Sahelian regions, its applicability to North America is controversial. There are no persuasive cases of actual "spreading of deserts" in North America, but certainly some of the associated effects are visible throughout the West and elsewhere: soil blowing, drying up of some landscapes owing to stream impoundment for domestic agricultural water use, loss of plant cover, and so on. While some of these things are to be found throughout North America, there seems little doubt that the consequences are accentuated in moisture-short regions, especially those with considerable solar radiation. The critical margin appears to be the less than twenty inch zone of rainfall. West of the line defining this zone (approximately the 100th meridian), one encounters increasing degradation of resources.

There is, for example, no way of farming shortgrass plains so as to avoid some soil erosion and loss of fertility and tilth. The effects can be moderated by careful methods: preserving trash cover on plowed fields, contour plowing to control water erosion, plowing plant refuse back into topsoil to rebuild tilth, and avoiding cultivation to the extent possible, especially during windy periods. New methods emerge almost yearly, as North Americans begin to understand ASAV environments. This is good, but the course of change still seems downhill: overgrazing and consequent destruction of forage cover continue; drastic disturbance of friable soils around the spreading urban centers of the Southwest is unrestricted; and chemical fertilizers continue to be used in large quantities in the Plains regardless of their destructive effects on tilth, including the deposition of harmful nitrates. Irrigation continues to salinize acreage and diminish water tables. Conservation and "new farming" groups insist on controls and changes: regrassing of high-risk cultivated areas, cessation of "soil mining" tactics of big grain farmers, control of destructive use of all-terrain vehicles, and so on. Drought accentuates the consequences of such uses, and the system of market agriculture does not allow for release of pressure during periods of drought. The new round of intensive grain production and cultivation that began in the 1950s may be setting up the region for another socioeconomic disaster, given another long drought period—for which the region is probably overdue.

Concerning the Amerind occupation, I noted that this had its dynamic qualities as well: it began with simple hunting bands and concluded with a major exploitative pastoral phase, with varying implications for physical resources. Although moisture levels in much of the Plains are certainly marginal at best for the growth of permanent tree cover, recent research suggests that deliberate firing of grass cover by Plains Indians to facilitate the bison harvest accentuated the treelessness.[22] Certainly trees always existed along watercourses: these "gallery forests" were crucial for both animal and human adaptation to the Plains for millennia, and the islanded forests in higher elevations like the Black Hills and the Cypress Hills were also important refuges for both humans and animals. Grass cover has been "managed" by humans at least since the onset of effective horse/bison pastoralism, and one might conclude that the contemporary emphasis on cattle in the drier portions is technically a continuation of the indigenous adaptations centered on bison. However, agriculture also was not foreign to the Plains in aboriginal times and was particularly prominent in the eastern half of the Plains, along the major rivers and streams. But there is no doubt that Indian adaptations of all types exerted much less pressure on resources than the contemporary commercial regimes.[23]

Let me illustrate the pressure on resources with a vignette: the events in Kiowa and Crowley counties, eastern Colorado—in the less than twenty inch rainfall area of the Great Plains.[24] In February 1977 high-velocity warm winds—a massive chinook—swept down from the Rockies and, in seven hours, caused a major dust storm that moved eastward from Colorado, eventually becoming visible over the mid-Atlantic. The fallout from the storm shadowed 248,000 square miles of the south-central United States. Most of the soil came from portions of eastern New Mexico and the two eastern Colorado counties, and in some districts as much as eleven inches of topsoil was removed from plowed fields.

Nearly the entire land surface of these two Colorado counties consists of friable soil that should be returned to grass and used as rangeland, but since the late 1960s, precisely the opposite has taken place: the soil has been used for grain production because the economics of livestock production and the costs of converting plowed land to range prevent individual farmers, however large in scale, from doing otherwise. One of the important economic factors is the federal disaster relief payment system, in which farmers simply turn over federal checks to their local banker to finance the loan payments for annual operations. This is part of what we might call the "Worster effect" (from Donald Worster's book [1979] on the dust bowl, where he suggests that federal programs for welfare relief and soil conservation since the 1930s, however well meant, have often created an unearned financial cushion, encouraging overexploitation of fragile, high-risk lands). *The crucial element here is institutional, not ecological;* that is, in a market and profit-making economy, agriculture is subject to forces outside the control of local resource users, who have to do what the general economic system requires of them, regardless of economic costs.

Is this "adaptation"? Of course it is, but adaptation of the producers only to the institutions, not to nature; or rather, not to the limits of safe exploitation, but only to some illusory ultimate potential. To include environmental values in the activity pattern, it is necessary to make them economically profitable so as to encourage choice of alternative, less abusive uses of the land. This will take some doing—something more than a well-meant conservation program. The problems will remain, and only a fundamental change in the institutions will solve them.

NOTES

1. Data on the effects of recent agricultural activities in the Great Plains are analyzed in McGinnies and Laycock (1988).

2. Webb's pioneer social-ecological analysis of the adaptation problem was done before the environmental movement had sharpened our perception of human irresponsibility and greed. Adaptation for Webb meant largely the fulfillment of human needs by overcoming the constraints of the environment.

3. Or stated another way, "adaptation" refers to the behavior of people confronted with changes in their milieu that require corresponding changes in established routine. The popular meanings of the term are specific: "getting along," "making do," "doing what you have to order to survive," "putting up with the situation as best one can," or "making the necessary changes." However, embedded in these terms is a dual meaning: *tolerating* or putting up with existing conditions and perhaps changing one's ways just enough to continue to exist; and making *drastic changes* in one's habits when the situation itself has changed. That is, adaptation as a form of behavior exists on a continuum from conformity to innovation. Studies of the use and management of physical resources, when accompanied by concepts from the adaptational lexicon, are likely to focus on specific empirical circumstances, such as the time of planting as related to weather cycles. Two anthropological case studies of agrarian situations utilizing the concept of adaptation effectively, but differently, are Bennett (1982) and Hanks (1972). In my own work, adaptation refers mainly to decision making in the uncertain physical and economic resource environment of market agriculture in the northern Great Plains. For the farmers in a Thai village in the completely different environment described by Hanks, adaptation refers to different and changing modes of raising rice and the effect these modes have on culture and social organization. However, both approaches refer to people adjusting to changing or fluctuating conditions. A collection of papers on Indian and Spanish American adaptations to the ASAV regions of North America is Knowlton (1964).

International conferences pertaining to arid lands have produced several volumes of proceedings, and the following are probably the best: UNESCO (1962); Hodge and Duisberg (1963); Dregne (1970); Whitehead et al. (1988). Perusing these works in order of publication can provide an idea of the expanding knowledge of aridity and human adaptation and, at the same time, the persistence of inappropriate methods of use. The symposia contain few papers on cultural and institutional aspects of arid lands adaptation, mainly because the study of aridity has been principally a specialty of natural scientists and some applied fields like irrigation agriculture.

4. There are a number of ways of classifying the differing environments or "biomes" in moisture-marginal portions of temperate lands. Shelford (1963) uses the following for North America: northern and southern temperate grasslands; the hot desert; a number of semidesert biomes; and a number of ecotones and "marginal contact" regions associated with deserts and grasslands and adjoining areas. The classification is complex; there are few hard boundaries. Data on human population and resources for arid lands also are difficult to interpret because of the ambiguous definition of the concept of aridity. Alan Eyre, of the College of the West Indies, in a paper presented at the International Arid Lands Conference in Tucson, Arizona, in 1985 (Eyre 1988), calculated the total number of inhabitants of "arid lands" of Asia, Europe, Africa, Australia, and the two Americas in 1980 as 412 million persons, the figure representing an 81 percent increase since 1960. About 62 percent of the 412 million were supported by "seasonal and irrigated agriculture," and 27 percent by "light industry, commerce, services": the remainder, 11 percent, lived by various activities, including pastoral nomadism, tourism, and minerals extraction. The figures should not be taken literally, but the trends are probably representative.

5. The paper also contains a good general review of human physiological adaptations to both humid and dry high-heat climates. See also the relevant chapters in Moran (1979).

6. This also leads some anthropologists to propose that hominids first evolved in the dry tropics. The idea is supported by the fact that the best prehominid fossil species are found in such an environment: southern and southeastern Africa. As Newman (1970) suggests, we are hairless, sweaty, thirsty mammals.

7. I refer here, and elsewhere in this chapter, to a current research program: Northern Plains Culture History Project, supported by the National Endowment for the Humanities.

8. The study of energy as a factor in social organization and change has attracted some attention among social scientists but has not become a major topic. The sociologist Leonard Cottrell (1955) and more recently Howard Odum (1970; 1983), have developed the issue in research on social and natural systems. In general, the current approach in the institutional social sciences is to fuse the energy concept with sys-

tems analysis. In anthropological studies of the culture-nature interchange in tribal societies, the analysis of energy has been identified with ecological research in which the main issue is how far tribal culture becomes part of natural ecosystems in the search for and production of food (Ellen 1982). Important concepts related to energetics theory are found in Lotka (1922, 147–51). Lotka was a biological ecologist, and his theories apply most cogently to aspects food and energy among animals and plants. Applications to human affairs are largely analogic and should be handled carefully.

9. Kroeber (1953) endeavored to correlate specific cultural types and patterns with various climatic and physiographic factors in an effort to put to rest a long-standing controversy over the nature of "culture areas" and their relation to the physical environment. He found some relationships, but no simple, one-to-one concordances. The Plains Indian/grasslands concordance was one of the best.

10. Webb (1932) was primarily concerned with the environmental differences between the West and the more humid East, and how eastern institutions would have to change in order to adapt. Turner (1958) was concerned with the "role of the frontier"—the bold individualistic spirit of the pioneer encountering a new world and how this pattern left its imprint on American culture. Webb underestimated the capacity of established institutions, such as land tenure, to handle new circumstances, for better or for worse; Turner neglected other forces in American society that were as influential as the "frontier" in shaping the American character. James C. Malin, writing a generation later, updated both writers, and added new insights (Malin 1984 is a recent collection of his essays).

11. The concepts discussed in this section are derived from "decision theory," a blend of economic behavior analysis and rational choice theory in social psychology. A standard treatment is Collingridge (1982). For an application of some of the concepts to farming strategies in the Great Plains, see Bennett (1982).

12. For an illuminating discussion of the variability and dynamism of western tribal economic and political culture, both before and after European contact, see Friesen (1984, chaps. 1–7).

13. Classic ethnographic descriptions of such peoples are found in Forde (1937). Forde's analysis of Shoshone adaptations was a pioneer ethnological statement of tribal people's ability to extract a living from even the most refractory environments. More recent ecological studies of human groups at the hunting-gathering level in dry environments provide a more dynamic analysis of coping behavior. See, for example, Turton (1977).

14. Considerable literature has accumulated on pastoralism owing to the efforts made by developing countries in Africa and the Middle East to sedentarize these peoples. For ethnographic description, one of the best works is Gulliver (1955). For

analyses of the pastoralism development problem in various parts of Africa, see Bennett, Lawry, and Riddell (1986).

15. Similarities and differences between Plains Indians and the rancher/cowboy population and its culture were noted by Webb (1932) and by Frantz and Choate (1966). For a comparison of rancher culture with that of pastoral nomads in the Old World, see Gilles (1987). Biographical accounts of the early range cattlemen are found in Atherton (1961).

16. See Lonsdale and Holmes (1981); cf. Parkes, Burnley, and Walker (1985). These studies feature the urban/rural dialogue: nucleated urban settlement is costly and reduces certain rural amenities but is necessary to provide desired facilities like higher education.

17. A modern study of the Sun Dance is Jorgensen (1972).

18. An introduction to locational theory is Haggett (1965).

19. See, for example, Royal Commission on Agriculture and Rural Life (1956). This report diagrams the location of towns and villages of various sizes in western Saskatchewan, showing patterning characteristic of Plains landscapes as predicted by locational theory. However, this pattern is the result of some eighty years of economic development that eliminated many early communities and permitted others to become larger.

20. There is, of course, an issue concerning Kraenzel's concept of "deficits" similar to that noted for Walter Webb: Kraenzel's idea that the inhabitants of the Plains deserve better is based on the assumption that, granted resources are refractory, correct adaptation will permit both human gratification and resource conservation. The possibility that people in regions like this, under our present economic system, must learn to live with less if they choose to remain in the region or if resources must be conserved, is difficult to acknowledge. The "out" is, of course, extensive government subsidy, which transfers the costs of the economic deficit onto more fortunate populations and regions. The problem is identical for the Third World ASAV countries. In other words, "social deficits" are inescapably rooted in resource deficits. Further difficulties in evaluating the social and economic position of populations in the drier parts of the United States arise when social indicators that define some measure of social well-being are compared with "hard" data on income, cultural facilities, and other signs of "progress" and urban amenities. David Smith (1973) used data on "social pathology" and "socioeconomic well-being" to create a series of scores for all forty-eight contiguous states expressing social well-being. The Great Plains and southwestern states split into two large groups: the north and central states had consistently high scores, some among the highest in the nation; the southern plains states had much lower scores, with some of the high scores being negative, like Texas with -536. That is, the index shows the effect of minority-group poverty populations

in depressing "well-being" in the South; conversely, the more egalitarian rural society of the northern and central states raises the index (Idaho has +606). Thus the "social deficits" of Kraenzal (1967) here seem to be reversed in the sense that, while sparse populations may limit access to certain facilities, it also may produce social harmony! Big cities are another factor: the low scores of Texas and Arizona are similar to those for certain eastern states with large cities, where crime, poverty, and so on, depress the figures. Americans have always had two images of rurality: limited and sullen, or free of conflict and replete with bucolic enjoyment. (For other discussions of cultural differences, see Gastil 1975).

Little research has taken place on the problems of urban environments in ASAV lands. Perhaps architects, occupied with the problems of hot, dry climate and energy-saving cooling methods, have done more than anyone else. For a review of current research on various urban aspects, see Whitehead et al. (1988, chap. 10, "Urban Environments"). Issues of domiciliary adaptation are in the long run less important ecologically than the demands that growing urban populations in ASAV lands place on the surrounding hinterland and its friable resources.

21. The following works provide an overview, with emphasis on social aspects, of the "desertification" literature and issues: Spooner and Mann (1982, esp. chap. 1); Council on Environmental Quality (1981); Dregne (1977). The map in Dregne's paper, which aims at showing the lands at risk, was originally drawn for the 1979 United Nations Conference on Desertification. Versions of the map have appeared in many other publications. The map is really only a cartography of ASAV lands, and not an index of desertification causes and effects.

22. A more recent study is Hildebrand and Scott (1987). For a recent archaeological analysis of aboriginal Plains inhabitants, see Caldwell, Schultz, and Stout (1983).

23. See the maps in chapter 1 of Petulla (1977) for distribution of rainfall, vegetation, and other physical features of the United States in the context of early settlement history. A recent general source of data on moisture regimes is Geraghty et al. (1973); cf. McGinnies and Laycock (1988).

24. This descriptive material is taken from Council on Environmental Quality (1981, 75–87).

REFERENCES

Atherton, Lewis. 1961. *The cattle kings*. Bloomington: University of Indiana Press.

Bennett, John W. 1969. *Northern plainsmen: Adaptive strategy and agrarian life*. Chicago: Aldine.

———. 1974. Anthropological contributions to the cultural ecology and management of water resources. In *Man and water: The social sciences in management of water resources*, ed. L. D. James. Lexington: University Press of Kentucky.

———. 1976. *The ecological transition: Cultural anthropology and human adaptation.* New York: Pergamon Press.

———. 1982. *Of time and the enterprise: North American family farm management in a context of resource marginality.* Minneapolis: University of Minnesota Press.

Bennett, John W., Steven W. Lawry, and James C. Riddell. 1986. *Land tenure and livestock development in sub-Saharan Africa.* AID Special Evaluation Study 39. Washington, D.C.: U.S. Agency for International Development.

Blouet, Brian W., and Frederick C. Luebke. 1979. *The Great Plains: Environment and culture.* Lincoln: University of Nebraska Press.

Caldwell, Warren W., C. Bertrand Schultz, and T. Mylan Stout. 1983. *Man and the changing environments in the Great Plains: A symposium.* Lincoln: Nebraska Academy of Sciences.

Collingridge, David. 1982. *Critical decision making: A new theory of social choice.* New York: St. Martin's Press.

Cottrell, Leonard. 1955. *Energy and society: Relations of energy, social change and economic development.* New York: McGraw-Hill.

Council on Environmental Quality. 1981. *Desertification of the United States.* Washington, D.C.: Council on Environmental Quality.

Cronon, William. 1983. *Changes in the land: Indians, colonists and the ecology of New England.* New York: Hill and Wang.

Dregne, Harold E. 1970. *Arid lands in transition.* Washington, D.C.: American Association for the Advancement of Science.

———. 1977. Desertification of arid lands. *Economic Geography* 53:325.

Ellen, Roy. 1982. *Environment, subsistence and system: The ecology of small-scale social formations.* New York: Cambridge University Press.

Ellison, Rosemary. 1980–83. The agriculture of Mesopotamia. *Tools and Tillage* 4:173–84.

Eyre, L. Alan. 1988. Population pressure on arid lands: Is it manageable? In *Arid lands: Today and tomorrow*, ed. Emily C. Whitehead et al., 989–96. Boulder, Colo.: Westview Press.

Forde, C. Daryl. 1937. *Habitat, economy and society: A geographical introduction to ethnology.* New York: Harcourt Brace.

Frantz, Joe B., and Julian Ernest Choate, Jr. 1966. *The American cowboy: The myth and the reality.* Norman: University of Oklahoma Press.

Friesen, Gerald. 1984. *The Canadian prairies: A history.* Lincoln: University of Nebraska Press.

Gastil, Raymond D. 1975. *Cultural regions of the United States*. Seattle: University of Washington Press.

Geraghty, James J., et al. 1973. *Water atlas of the United States*. Port Washington, N.Y.: Water Information Center.

Gilles, Jere L. 1987. *Nomads, ranchers, and townsmen: Sociocultural dimensions of pastoralism today*. American Association for the Advancement of Science Symposium. Boulder, Colo.: Westview Press.

Grigg, David. 1980. *Population growth and agrarian change*. Cambridge: Cambridge University Press.

Gulliver, P. H. 1955. *The family herds: A study of two pastoral tribes in East Africa, the Jie and Turkana*, London: Routledge and Kegan Paul.

Haggett, Peter. 1965. *Locational analysis in human geography*. London: Arnold.

Hanks, Lucien. 1972. *Rice and man: Agricultural ecology in Southeast Asia*. Chicago: Aldine Atherton.

Hildebrand, David V., and Geoffrey A. J. Scott. 1987. Relationships between moisture deficiency and the amount of tree cover on the pre-agricultural Canadian plains. *Prairie Forum* 12:203–16.

Hodge, Carle, and Peter C. Duisberg, eds. 1963. *Aridity and man: The challenge of arid lands in the United States*. Washington, D.C.: American Association for the Advancement of Science.

Jorgenson, Joseph P. 1972. *The sun dance religion*. Chicago: University of Chicago Press.

Khazanov, A. M. 1978. Characteristic features of nomadic communities in the Eurasian steppes. In *The nomadic alternative*. The Hague: Mouton.

———. 1981. Myths and paradoxes of nomadism. *Archives Européenes de Sociologie* 22:141–53.

Kirk, R. L. 1981. *Aboriginal man adapting: The human biology of Australian Aborigines*. New York: Oxford University Press.

Knowlton, Clark S., ed. 1964. *Indian and Spanish American adjustments to arid and semiarid environments*. Contribution 7, Committee on Desert and Arid Zone Research. Lubbock: Texas Technological College.

Kraenzel, Carl. 1967. Deficit creating influences for role performance and status acquisition in sparsely populated regions of the United States. In *Symposium on the Great Plains of North America*, ed. C. C. Zimmerman and S. Russell. Fargo: North Dakota Institute for Regional Studies.

Kroeber, Alfred L. 1953. *Cultural and natural areas of native North America*. University of California Publications in American Archaeology and Ethnology 38. Berkeley: University of California Press.

Lonsdale, Richard E., and John H. Holmes. 1981. *Settlement systems in sparsely popu-*

lated regions: The United States and Australia. London: Pergamon Press.

Lotka, Alfred J. 1922. Contribution to the energetics of evolution. *Proceedings of the National Academy of Sciences* 8:147–51.

McGinnies, William G., and William A. Laycock. 1988. The Great American Desert—Perceptions of pioneers, the dustbowl, and the new sodbusters. In *Arid lands: Today and tomorrow*, ed. Emily C. Whitehead et al., 1247–54. Boulder, Colo.: Westview Press.

Malin, James C. 1984. *History and ecology: Studies of the grassland.* Ed. R. P. Swierenga. Lincoln: University of Nebraska Press.

Moran, Emilio F. 1979. *Human adaptibility: An introduction to ecological anthropology.* North Scituate, Mass.: Duxbury Press.

Newman, Russell W. 1970. Why is man such a sweaty and thirsty animal: A speculative review. *Human Biology* 42:12–27.

Odum, Howard T. 1970. *Environment, power, and society.* New York: Willey-Interscience.

———. 1983. *Systems ecology: An introduction.* New York: Wiley-Interscience.

Palliser, John. 1859. *Papers relative to the exploration by Captain Palliser of that portion of British North America which lies between the northern branch of the river Saskatchewan; and the frontier of the U.S.; and between the Red River and the Rocky Mountains.* Presented to both Houses of Parliament by Command of Her Majesty. London, June, 1859. Facsimile reprinting by Greenwood Press, N. Y., 1969.

Parkes, D. N., I. H. Burnley, and S. R. Walker. 1985. *Arid zone settlement in Australia: A focus on Alice Springs.* Tokyo: United Nations University.

Petulla, Joseph M. 1977. *American environmental history: The exploitation and conservation of natural resources.* San Francisco: Boyd and Fraser.

Powell, John W. 1962. *Report on the lands of the arid region of the United States.* Ed. Wallace Stegner. Cambridge, Mass.: Belknap Press.

Royal Commission on Agriculture and Rural Life. 1956. *Service centers.* Rural Life Report 12. Regina: Government of Saskatchewan.

Shelford, Victor W. 1963. *The ecology of North America.* Urbana: University of Illinois Press.

Smith, Henry Nash. 1950. *Virgin land: The American West as symbol and myth.* Cambridge: Harvard University Press.

Spooner, Brian, and H. S. Mann, eds. 1982. *Desertification and development: Dryland ecology in social perspective.* New York: Academic Press.

Stegner, Wallace. 1962. *Wolf Willow: A history, a story and a memory of the last Plains frontier.* New York: Viking Press.

Turner, Frederick Jackson. 1958. *The frontier in American history.* New York: Henry Holt.

Turton, D. 1977. Response to drought: The Mursi of southwestern Ethiopia. In *Human ecology in the tropics*, ed. J. P. Garlick and R. W. J. Keay. New York: Halsted Press.

UNESCO. 1962. *The problems of the arid zone*. Proceedings of the Paris Symposium. Paris: Arid Zone Research.

Webb, Walter P. 1932. *The Great Plains*. New York: Ginn.

Wells, Philip V. 1969. Postglacial vegetational history of the Plains. *Science* 167:1574–81.

Whitehead, Emily C., Charles E. Hutchinson, Barbara N. Timmerman, and Robert F. Varady. 1988. *Arid lands: today and tomorrow*. Boulder, Colo.: Westview Press.

Worster, Donald 1979 *Dust bowl: The southern plains in the 1930s*. New York: Oxford University Press.

Chapter Two

Pastoral Nomads in the Past, Present, and Future: A Comparative View

Anatoly Khazanov

Were I to put my ideas about pastoral nomads in a nutshell, I would express them as follows: Their past was unique, their present is transitional (even miserable, if we compare it with their political influence in the past and their former role in world history), and as for the future—they have no future at all, if present tendencies continue.

From the economic point of view pastoral nomadism may be defined as a distinct type of food-producing economy in which extensive mobile pastoralism is a predominant activity; the majority of the population is drawn into periodic migrations caused by a necessity to maintain herds all year round on a system of free-range grazing. Unlike other forms of pastoralism, pure pastoral nomadism lacks agriculture even in a supplementary capacity. This definition implies two oppositions: between pastoralism and agriculture and between nomadism and sedentarism. In my opinion nomadism is a specific kind of mobility, distinct from wandering and migration (for further details, see Khazanov 1984, 15 ff.). In this respect pastoral nomads are quite different from the early nineteenth-century American Indians of the Great Plains, the mounted hunters of bison, although they shared some characteristics.

We need not discuss at length the essence of the Great Plains Indian cultures, but one may debate whether these people should be considered hunters or pastoral nomads. Many scholars do agree now about their prime characteristics.

1. Like pastoralist cultures, they developed and functioned in an arid environment unsuitable for primitive horticulture (Webb 1931), and in both cases the availability and the use of animals for transport facilitated their development of vast territories.

2. They did not originate from indigenous and spontaneous development, or even from ordinary diffusion, but grew out of early colonialism. Plains Indians cultures were created by indigenous groups that, pressured by white Euro-Americans, had to abandon or transform their previous economies, including horticulture, and readjust to the new natural and political environment (Ewers 1955, 152; Wedel 1961, 284–92), a readjustment aided by horses and guns borrowed from the whites.

3. Except for the dog, the horse was the only domestic animal kept by the Plains Indians, whereas most pastoral nomads kept many species; in this respect the Plains Indians were also quite different from the true Indian pastoralists, the Navajos (Downs 1964). Further, the Plains tribes did not use horse products as food, or at least used them very little. The backbone of the Plains Indians subsistence economy was bison hunting of a type different from pedestrian big game hunting, and it lacked analogues in most other hunter cultures (Murdock 1968, 13–15). One of the few, incomplete exceptions is the hunting practiced by some peoples of the taiga and tundra zones of Eurasia, who kept domestic reindeer primarily for transport, while their economies over the centuries remained basically food-extracting ones (Khazanov 1984, 112–13).

4. Although the horse was not directly used for food by the Plains Indians, it made more intense hunting possible and thus provided a means for obtaining sustenance. The very presence of the horse introduced some pastoralist elements into the Plains economy, since domestic animals must be looked after, pastured, provided with fodder, and so on.

5. Though the horse in the Plains Indians cultures acquired a cultural and social value that had analogues in pastoral cultures, new social stratifications and concomitant institutions, such as militarism, may not be ascribable to its introduction alone (cf. Ewers 1955). The general historical circumstances in the area, including various influences from the white Euro-Americans, must also be taken into account.

Whether one calls the Plains Indians hunter-gatherers or nomadic pastoralists largely depends on one's concept of pastoralism. If one agrees with Oliver (1962b, 35) that pastoralism is a sociocultural system, a lifeway, then one will justly pay particular attention to its similarity to the mounted hunting of the Great Plains. However, to me pastoralism is a distinct type (or types) of food-producing subsistence economy, different from mounted hunting. Plains Indians were not typical hunters; nevertheless they remained hunters, and the general orientation of their *food-extracting* economy was different from that of the pastoralist *food producer*. Correspondingly, the character of periodic movements of the hunter-gatherer and pastoral nomadic peoples, their underlying reasons for living as they did, and the determinants of their annual economic cycle were also quite different (Oliver 1962a). Even similar cultural elements in these two kinds of societies sometimes had different underlying reasons. For example, whereas the Plains Indians deliberately burned the grass to help with the bison harvest, the nomads of the Eurasian steppes used this practice not so much for hunting as to increase pasture for domestic stock. For all these reasons any proposal to speak of the Great Plains cultures as practicing a species of pastoralism seems to me an overestimation of a pastoral side of cultures that basically hunted to survive.

Other differences between nomadic and Plains Indian societies are important, for pastoral nomadic societies have always depended on their sedentary counterparts because of some basic weaknesses. Though pastoral nomadism in the past was suitable to the vast territories in the arid, tundra, and alpine zones of the Old World, this specialized system had many shortcomings. Its ecological and economic foundations made it incapable of long-term stable economic growth based on technological innovations and increased productivity. Because the system depends on a delicate, unstable balance among the natural resources used by livestock, the number of livestock, and the population size, it is homeostatic, although its regulation takes the form of cyclical fluctuations. Long-term maximization of the number of livestock requires increasing the production base through territorial expansion. Periodic droughts, epizootics, and other natural calamities strongly limit long-term livestock reproduction on an expanded scale, and therefore the pure pastoral nomadic economy is doomed to stagnation.

Furthermore, pastoral nomadism, unlike hunting-gathering, was in no way in the mainstream of human social evolution but rather was a by-product of two revolutions: a Neolithic one and an urban one, in Gordon Childe's terminology (or a transition from primitive to traditional societies or, in still other terms, the origin of the state). The necessary prerequisites for pastoral nomadism were first created by the transition from food-extracting to food-producing economies usually labeled the Neolithic revolution: that is, the beginning of agriculture and animal husbandry and the potential for their specialization in the ecological zones most suitable for them. Some scholars still hold that in several areas pastoralism, or even pastoral nomadism, might have developed out of hunting, particularly the hunting of herbivorous mammals, and some continue to insist that pastoralism emerged earlier than agriculture. However, only advanced gatherers had created the conditions necessary for domestication of livestock: a relatively sedentary way of life and a surplus of vegetable products for feed.

The mobile life of hunters and their lack of necessary fodder reserves were serious obstacles to the domestication of individual animals, let alone entire herds. They could only borrow already domesticated animals: for example, the Navajos who after the Pueblo uprising of 1680 turned to pastoralism (Underhill 1956, 41–43; Vogt 1961, 296; Ellis 1974, 309–24, 481). Though this was one way of disseminating animal husbandry and pastoralism, it was not the main one.

Although the Neolithic revolution created the conditions necessary for the emergence of pastoral nomadism, these conditions were not fully realized for many millennia. Real pastoral nomadism, with both mounted herders and a complete separation from agriculture, emerged late, only at the turn of the second millennium B.C., even in the Near East and in the Eurasian steppes—the main centers of its dissemination. In both areas, it developed from a mixed economy through intermediate forms of extensive pastoralism, and the immediate transition to it was linked to specific climatic changes. In other areas nomadism was formed or spread even later, under the direct or indirect influence of existing forms of pastoralism and pastoral nomadism.

From the beginning of the Neolithic revolution and for many millennia afterward, food-producing economies in the Old World were usually based on a combination of agriculture and animal husbandry. I have expressed the opinion that in the Neolithic, and even in the Bronze Age, pastoral nomadism did not exist (Khazanov 1984, 85 ff.). This opinion still seems to me the most plausible. Horses and camels for riding appeared on a large scale only in the first millennium B.C., and it is also true that developed forms of pastoral nomadism are inseparable from riding animals. However, we also know of other types of pastoral nomadism, in eastern Africa and northern Asia, that lack animals for riding. Inconclusive as most of the archaeological data are, they leave room for some pastoral nomadism—for example, the Neolithic Sahara or in several regions of the Eurasian steppes in the early Bronze Age. A kind of primitive pastoral nomadism also might have been practiced in the fifth to the fourth millennium B.C. by the former hunters-gatherers of the Sinai who, turning to other economic activities because of the climatic desiccation, borrowed stock (goats) from their neighbor agriculturalists (Bar-Yosef, pers. comm.). But these incipient forms, if they existed, were short-lived and had no direct historical genetic relation to those forms that emerged from the turn of the second millennium B.C. One may say that these incipient forms failed, and the reasons for their failure provide us with a better understanding of the essence of pastoral nomadic societies.

Pastoral nomadism had many disadvantages compared with mixed economies. It was a perfect specialization for assimilating certain ecological zones, but this very specialization made it not a self-sufficient economic system. It depended for its normal functioning upon other economies just as societies based on pastoral nomadism depended upon other generally sedentary societies. Its location in arid geographical zones and its mobile way of life left little room for agriculture and crafts—though written sources since the very first mention of nomads make it clear that agricultural products formed an important part of their dietary systems. These sources, as well as numerous archaeological data, demonstrate beyond doubt that nomads procured a substantial part of their material culture from sedentary territories, as well as other aspects of culture and even ideologies. As the nomadic economy needed to be supplemented with agricultural products and crafts from external sources, so nomadic culture needed sedentary culture as a source—a component and a model for comparison, rejection, and imitation. Another characteristic of pastoral nomadism is its constant instability. Ecology and biology made periodic losses of livestock inevitable because of various natural calamities, and perhaps also because of some other still uncomprehended regulatory mechanisms operating in herds of domesticated ungulates. On the other hand, stock numbers sometimes outgrew the carrying capacities of available pastures. It was just such cyclical fluctuations that maintained a long-term balance in the pastoral nomadic economy, however ruinous they might be in the short run. For that reason, too, nomads needed sedentary societies as a reserve fund vital for their survival, and therefore pastoral nomadic societies had to adapt not only to specific natural environments but also to external sociopolitical and cultural environments.

Given these disadvantages, pastoral nomadism required a specific motivating stimulus to come into existence—generally climatic changes that modified natural environments and led to a crisis that required the reorganization of former economies. For example, the desiccation of the climate in the second half of the second millennium and the beginning of the first millennium B.C. served as a major impulse for a transfer to pure nomadic forms of pastoralism in the Near East and Eurasian steppes (Khazanov 1984, 90–102). In contrast to the earlier short-lived and abortive experiments, the pastoral nomadism that emerged at the turn of the second millennium B.C., and in some regions even later, turned out to be viable for long historical periods and sometimes even spread into a number of marginal ecological zones. Why?

The reasons are not only ecological and economic but also sociopolitical. Beginning from the first millennium B.C., pastoral nomadism was emerging and functioning in a most favorable milieu because of the existence of developed sedentary societies with which nomads-to-be could maintain close economic, political, and cultural ties from the very outset. In contrast, the societies of the Plains Indians could not be said to have relied culturally for much of their short history on sedentary societies, though they were influenced by the prairie Indian cultures and adopted some of their technology and culture and, occasionally, aspects of their sociopolitical organization. Until late in the white incursion, they did not absolutely require outside trade items, though they might value them (see below). On the other hand, societies based on pastoral nomadism can never be independent and closed economic, sociopolitical, or even cultural systems because the specialized pastoralist economy itself cannot provide all its own requirements. However for its most efficient functioning, it needs sedentary societies of a certain developmental level—not primitive and not industrial, but traditional ones—especially traditional state societies such as the ancient and medieval states of the Near and Middle East, Central Asia, or China.

Unlike primitive societies, traditional state societies provided the best possibilities for the development and long-term historical functioning of pastoral nomadic societies. Traditional state societies created a surplus product and mechanisms for its extraction, distribution, and redistribution; they had stratified social systems, a division of labor, exchange and trade, and a culture, including the "Great Tradition," on which nomadic cultures might depend. The pastoral nomadic societies were themselves, with rare exceptions, only a specific variety of neighboring traditional societies.

But nomads had some military and social advantages in their interrelations with their sedentary counterparts. Up to modern times, these advantages often allowed them to transfer these interrelations from the purely economic onto a political plane. In other words, their military superiority gave them the leverage of political domination. That was particularly true for the Great Nomads of the Near and Middle East and the Eurasian steppes. In these militaristic societies every male was a warrior, most of them mounted, capable of the intrusions and conquests so numerous in ancient and medieval history. Economic and social backwardness may turn out to confer military and political advantage.

The situation changed as increasing numbers of traditional societies were transformed into industrial ones while the nomads did not, and could not, undergo substantial changes. They first lost their military superiority and political independence and afterward were incorporated into developed or developing states, sometimes with results as disastrous to them as the conquest of the Plains was to the Plains Indians. Eisenstadt (1983) was completely right when he stated that "the mere destruction of traditional forms did not necessarily assure the development of a new, viable, modern society, and . . . very often the mere disruption of traditional settings—be they family, community or even sometimes political settings—tended to lead to disorganization, delinquency, and chaos rather than to the setting up of a viable modern order."

This process has affected pastoralist peoples from Central Asia, Africa, and other areas as much as it has affected Native American mounted hunters, as European-based cultures have surrounded and conquered them. I will illustrate this process through the example of the Kazakhs, in the past the most numerous nomadic people in the Eurasian steppes, and the people who retained the pastoral nomadic way of life longer than most of the other peoples of the USSR. Their example may be of particular interest to students of the American Plains, since as one scholar stated, "In many ways the advance of Russia into the Kazakh Steppe paralleled the movement of the American Frontier into the Indian country of the Central Lowland and Great Plains of the United States." (Demko 1969, 33).

The steppes in the temperate zone of Eurasia constitute a wide belt stretching all along the way from Romania to the Altai Mountains and then, with some breaks, up to northern China. In the north, steppes change to wooded steppes. In the south, in the lower Trans-Volga areas and in Central Asia, steppes give way to semideserts and deserts, among which there are great ones, as Kyzyl-Kum, Kara-Kum, and the Gobi. Pastoral nomadism here emerged in the central steppe areas crucial to the maintenance of a nomadic economy, the main habitation zone for most nomads for almost three thousand years. Though the wooded steppes were used by some groups of nomads in summer, in winter they were useless because the season was too long and snows too thick for the livestock to feed. On the other hand, not all steppe nomads roamed to semideserts and especially to deserts, even for the winters, since the pastures in deserts were far less productive than the steppe ones and herding one's sheep required 1 hectare (2.47 acres) in steppes and 3 to 6 hectares (7.4 to 12.8 acres), sometimes even 10 hectares (24.7 acres), in deserts. Living in the temperate zone, where seasonal and even interseasonal fluctuations of temperature, rain and snowfall, and vegetation were great, the nomads in the Eurasian steppes practiced a clear division of pastures for winter and summer and even spring and autumn that resulted in long-distance pastoral migrations, sometimes reaching a 1,000 or even 1,500 kilometers.

The migratory routes were fairly fixed in space and time, and times to change pastures were usually observed very strictly, for pastures were to be used when they reached their peaks of productivity or were better than other pastures in a given season. Pastoral migrations required careful timing, considering the problem of feeding and watering the stock—usually they occurred when the grass on the way was fresh. If grasses dried out, some areas en route became hardly passable. When the Kazakhs had to break these rules of annual migrations, they were always confronted with heavy losses of stock. Hence the subsistence territory of the Kazakhs included not only seasonal pastures, sometimes for different animals of different species, but also stable migratory routes (Masanov 1984).

In the second half of the eighteenth century and the first half of the nineteenth, Kazakhstan was subdued and annexed by a Russia that was beginning to industrialize. The impact of colonialism on the Kazakhs was identical to its effects on the pastoralists of Africa and on the Plains Indians: loss of land, restriction of movement, declining living standards, and degradation of the natural environment (cf. Silitshena's chapter in this volume). The government made the Kazakhs register in fixed counties, prescribed their migratory routes, and declared all the Kazakh lands the property of the state. It also took away the Kazakhs' summer pastures, and sometimes even winter quarters, to settle them first with Cossacks and then with Russian peasants. About 1.5 million new colonists from European Russia came to Kazakhstan at the end of the nineteenth century and the beginning of the twentieth (Dakhshleger 1965, 51), and these new settlements sometimes blocked migratory routes. At the same time, the imposition of a money tax, and the necessity to buy products of agriculture and industry under unfavorable conditions, pushed the Kazakhs toward dependence on external markets and overstressed their traditional subsistence-oriented pastoral economy—exactly what happened to the Plains Indians as the fur and buffalo robe trades expanded.

Kazakh pastoral nomadism was gradually ousted to the arid areas of central and southern Kazakhstan, resulting in an overgrazing of the pastures the Kazakhs still possessed and in their impoverishment. Many had to migrate to other regions, settle on the land and cultivate crops, or even work for new colonists. By the time of the revolution, their traditional pastoral nomadic economy survived only in the central and southern parts of the country. No wonder conflict between indigenous populations and the Slav settlers became acute.

During the whole nineteenth century the Kazakh population in Kazakhstan grew very slowly (Bekmakhanova 1980, 194), and between 1902 and 1913 it diminished by 9 percent. Then followed the uprising of 1916, the turmoil years of the revolution, and civil war, during which the Kazakhs suffered much from both sides (the Whites and the Reds), and the starvation of 1921–22. All these events cost the Kazakhs more than million souls (Pipes 1964, 172–75).

During the 1920s the Kazakh pastoralist economy managed to recover, because some of the pastures were taken away from the Russian settlers and returned to nomads and seminomads. Then in the 1930s came the traumatic

Table 2.1 Average Number of Livestock per Household in Kazakhstan

1870	1881	1885	1890	1900	1910	1915	1920	1922	1923	1925	1926	1929
59	29.9	27.0	23.3	22.3	21.2	26.0	14.8	9.0	10.1	24.4	33.3	43.9

Sources: Ashmarin 1925, 123; Desiat let Kazakstana 1930, 208.

events of forced collectivization and the bloody settlement of Kazakh nomads on fixed lands. In a few years, about 550,000 nomadic and semi
nomadic households were forced to settle, many in waterless regions where
not only agriculture but even pastoralism was impossible; others were
moved to towns and cities to turn them into industrial workers, but they
could not find even unskilled work (Nurmukhamedov, Savosko, and Suleimenov 1966, 195–96).

The forced denomadization and collectivization of the Kazakhs met with
widespread opposition. Many slaughtered their stock or tried to drive it
into China. Those who resisted were killed or deported if they did not manage to migrate abroad. All this cost the Kazakhs again more than a million
lives and decimated their herds. However, the political aims were achieved:
the traditional nomadic way of life ceased to exist; the Kazakhs were "domesticated" and became a minority in their own country (Allworth 1967,
103), as the Plains Indians had half a century earlier. Pastoralism in the Kazakh steppes was almost destroyed, and it took many years for the livestock
even to approximate the prerevolutionary level.

Even under socialism of the type practiced by the communist countries,
the consequences of modernization were not everywhere as full of blood-
shed and as disastrous to the nomads as they were for the Kazakhs. An attempt to sedentize the Mongols undertaken in the 1930s failed. Because of
this failure, the Mongol government that imitated the Soviets was taught a
good lesson and had to conclude that the collectivization of the nomads and
their sedentarization were different things. The collectivization in Mon-
golia in the 1950s was accomplished without human losses, and even at the
end of the 1970s 490,000 Mongols remained pastoral nomads, while
300,000 were involved in sedentary agriculture and 240,000 were town
dwellers (Graivoronsky 1979).

Table 2.2. The Kazakh Population in Kazakhstan

Year	Kazakh Population	Kazakhs as Percentage of Total Population
1830	1,300,000	96.4
1850	1,502,000	91.1
1860	1,644,000	
1870	2,417,000	
1897	3,000,000	79.8
1926	3,713,000	57.1
1939	2,640,000	38.2
1959	2,755,000	30.0
1979	5,289,000	36.0

Sources: Bekmakhanova 1980, table 28; *Aziatskaia Rossiia* 1914, 82; censuses of 1926, 1939, 1959, 1979.

Table 2.3. Amount of Livestock in Kazakhstan

Year	Head
1913	29,931,700
1916	28,499,400
1920	16,300,300
1928	30,350,900
1933	4,800,600
1941	12,490,000
1946	14,724,100
1951	23,973,400
1959*	32,736,000
1965**	29,500,000
1970**	29,400,000
1975	29,700,000

Sources: Desiat let Kazakstana 1930, 206; Nurmukhamedov, Savosko, and Suleimenov 1966, table 37; Neishtadt 1960, 30, 56, 117, 210; *Ekonomicheskoe razvitie Kazakhstana* 1978, 107.
*Without camels.
**Without camels or horses.

The patterns of change that affect various pastoral nomads in the modern world are different, but their consequences are similar: the disappearance of traditional forms as modern economic systems or supersystems are imposed. A pastoral nomadic economy was always specialized and could be nothing else; however, a traditional nomadic specialization was natural and organic, more an ecological than an economic specialization. In this regard pastoral nomadic adaptations to the arid environment parallel Plains Indian adaptations before the development of the fur trade. Both responded to ecosystems first and to markets only secondarily. In the past, pastoral nomadic economies were never deliberately profit oriented or consistently aimed to meet market demands, since among many other things it was difficult for such an economy to maintain a high level of productivity. Production in general, and particularly production for sale, was subject to great annual fluctuation. Hence the pure nomads preferred to overcome the deficiencies of their specialization by noneconomic political means, that is, maintaining asymmetrical relationships with the sedentaries (Khazanov 1984). When subsistence-oriented economies, even specialized ones, are pushed to produce a regular and increasing surplus in accordance with external market demands—by traditional technologies or by technologies that have undergone some kind of modernization but still operate within the traditional social organization—they become overstressed. They cannot compete with or even adapt to the modern economies, and destructive social consequences are inevitable.

Pastoral nomadism in its traditional and pure forms—in which it existed for two or even three thousand years, and the remnants of which some of us can witness in our fieldwork—is dying, if not already dead. It is dying because it has proved less compatible with industrial society than with primitive society. As traditional societies transform themselves into modern or modernizing ones or are pulled into the orbit of such societies, traditional pastoral nomadic societies are also changing. I do not see any long-standing tendencies that can seriously slow this process, or even more turn it back, although a radical transformation from one way of life to another is always painful and often not very successful from either an economic or a social point of view. Therefore the problem consists in finding the kind of modernization that is the most suitable for traditional pastoral nomadism.

Many governments and government experts, particularly in the Second and Third Worlds, favored and still favor sedentarization as almost the only alternative to pastoral nomadism, which they consider at best antiquated and unprofitable and at worse an obstacle to progress. Not infrequently this opinion is inspired by underlying political considerations—a desire to impose upon the nomads the power of the central government. However, transforming nomads and pastoralists into sedentary cultivators generally means that vast desert and semidesert territories unsuitable for cultivation are wasted and cease to be used for food production.

One may also wonder whether sedentarization is always desirable and expedient for the nomads themselves. Though the overtly bloody enforced forms of sedentarization practiced in the first half of this century by the governments of the USSR and Iran have not become a general practice, even now sedentarization sometimes occurs under external pressures, after the nomads are impoverished. Under such conditions, it can hardly be considered free. In the past mass sedentarization has been successful only when nomads migrated into areas favorable for agriculture and often seized by force arable lands in oases or in zones of dry agriculture. But at present sedentarization does not take this form, and it faces additional difficulties because of a shortage of land suitable for cultivation, demographic pressures, and so forth. Former nomads and pastoralists often have to settle in marginal areas and cultivate lands that agriculturalists themselves consider of little use. Not infrequently these lands quickly become degraded because of overcultivation (see, for example, Silitshena's chapter in this volume). The effective development of even some of these lands at best depends on expensive irrigation projects and other large-scale capital investments. With rare exceptions, so far neither national governments nor international agencies are in a hurry to make such investments. All in all, I can only agree with Salzman (1980, vii): "Sedentarization, viewed as an inevitable and necessary step in furthering progress and advancing civilization, and pressed upon nomadic peoples by external forces, can have detrimental consequences not only for the nomadic peoples themselves but for the large societies of which they are part."

Many Western experts advocate commercial livestock production and ranching as another way to modernize subsistence pastoralism. However, as other essays in this book suggest, ranching and irrigated intensive production systems have often failed or become excessively costly. This development, hardly possible in the deserts and semideserts for environmental reasons, is difficult even in less arid areas. Besides, administrators and planners have not taken into account the peculiarities of the social organization and landownership of the pastoral nomads and have forgotten in particular that in traditional nomadic societies land had no market value and was not sold.

The open range system that developed in the western half of North America in the last quarter of the nineteenth century and the first half of the twentieth was from the outset different from the traditional pastoral nomadic economies in its land tenure system and its operation within a framework of a capitalist profit-oriented economy. As Bennett notes, "Livestock was produced for sale from the very beginning; there was no introductory period of subsistence production, as there was in the Middle East, Central Asia, and Africa. . . . North American cattlemen, despite their romantic traditions, were businessmen from the beginning" (Bennett 1985, 90). All this provided the open range system with irrigated pastures, machinery, tame forage, breeding, fences, and so on.

Traditional nomads and extensive pastoralists could not be turned into ranch owners without drastic changes in their social organization connected with a privatization of communal forms of land tenure and landownership (see, for example, the chapter by Bekure and Ole Pasha in this volume on the Kenya Maasai case). Under conditions now prevailing in the Third World countries, this privatization inevitably increases the number of displaced and unemployed persons. Apart from the moral issue, one may wonder whether other branches of modernizing economies are capable of providing these people with work and possibilities for readjustment. Are they doomed to join the ranks of the lumpens?

What kind of economy must substitute for the pastoral nomadism in arid environments, and is self-sufficiency possible there at all? In earlier publications I have shown that for its normal and successful functioning traditional pastoral nomadism depended on the external resources provided by the sedentary world (Khazanov 1981, 1984). In his chapter in this volume, Bennett argues that the relative stability of modern economies in the arid or semiarid environment is also connected with external economic and financial input. However, even today extensive pastoralism in many arid areas remains less labor intensive and more profitable than many other forms of economy. Pastoralism originally developed in just those regions where agriculture was impossible or economically less profitable, and in most of them the situation remains basically the same. For example, in Central Asia husbandry in which herds are pastured all year round is 2.5 times as profitable as cotton growing (Fedorovich 1972, 218). Extensive pastoralism remains viable also because it is less specialized than pure pastoral nomadism and is capable of combining stockkeeping with other economic activities, including agriculture and industry.

In addition, there is another problem connected with the conservation of nature. We can hardly expect any positive and long-range results until solutions confront the people and their subsistence problems. The opinion that extensive pastoralism necessarily destroys the environment is unfounded. Extensive pastoralism in general, and pastoral nomadism in particular, resulted in only one essential ecological change: wild herbivores were gradually replaced by domestic stock. The extermination and ousting of the wild species, however, was not senseless and predatory, since they competed for pastures with domesticated animals. Pastoralism was a much more efficient economic system than hunting, procuring not only meat but also milk and milk products, blood, wool, and means of transport.

In transforming the species system, pastoralists were much more moderate than agriculturalists; 75 million bison were exterminated in the Great Plains in no more than 157 years between 1730 and 1887, but mostly only after conquest in such areas as Indian Territory and Pine Ridge. In the Eurasian steppes a similar process took 3,000 years and was particularly intense only during the last few hundred. Herds of gazelles (*Gazella subgutturosa*) and antelope (*Saiga tatarica*) survive to the present day in several remote regions of the Eurasian steppes where agriculture and industry did not gain the upper hand over pastoralism. That proves the point. Neither mounted hunters nor mounted pastoralists were the major exterminators of wild fauna in the Plains, steppes, and other dry areas. Agriculturalists were. Generally pastoralism does not lead to a deterioration of flora either, and moderate grazing may even be necessary to the normal functioning of steppe vegetation, since stock trample seeds of wild plants into the soil, eliminate weeds alien to local flora, and manure the soil. Under pastoralism the destruction of the environment took place only in the cases of overgrazing, and steppe vegetation can regenerate as long as overgrazing is temporary, as it is in traditional forms of pastoral nomadism. In modern times overgrazing in many parts of the world is connected not so much with cycles of pasture abuse and recovery as with political factors: alienation of the lands belonging to the nomads and restriction of their movements increases the pressure on the pastures remaining in their possession.

Recent bold, large-scale attempts to change semiarid regions drastically have often failed or had disastrous repercussions. The Russian colonists in Kazakhstan practiced extensive dry farming based on grain monoculture without fertilizers, and even before the revolution this had led to soil exhaustion and diminishing grain yields (Dvorkin and Maslov 1971, 36, 91–92). In consequence, they started to shift toward a more balanced grain/livestock economy and dairy farming (Demko 1969, 172), but the lesson of diversification was soon forgotten, as was the American experience of dust bowls (although the environmental and climatic conditions of northern Kazakhstan are similar to those of the American Plains, with their favorable and unfavorable cycles for grain agriculture). During the "virgin lands" campaign started in the 1950s and aimed at sowing wheat on huge tracts of land in the northern Kazakhstan steppes, incorrect agrotechnology brought in weeds as well as wind, soil, and water erosion, and in less than fifteen years generated 3 million hectares (7.5 million acres) of sand and made about 12 million hectares (30 million acres) of land liable to wind erosion (Uteshev and Semenov 1967, 5; McCauley 1970; Komarov 1978, 53).

In many countries the overexploitation of productive ecosystems in the arid zones, through imposed commercialization and attempts to adapt and apply modern technologies, gives rise to desertification and degradation of vegetation, soil, and water (Reining 1978; Galaty, Aronson, and Salzman 1981). Some of the schemes have failed because the desire to intensify production was not accompanied by the creation of proper infrastructures. The loss of control by pastoralists over the stock they were herding—its alienation from the immediate producers and concentration in the hands of absentee owners—may also contribute to a serious ecological and social disequilibrium in places such as the Sahelian zones of West Africa (Maliki 1986, 4–5). Intensifying extensive pastoralism and applying scientific methods turned out to be difficult even when the planners did not want to turn the pastoralists into capitalist ranch owners. Attempts to increase the productivity of pastures and improve breeds while maintaining the traditional pastoralist way of life often had undesirable consequences, unexpected by the planners, particularly when the problem of balance was ignored. Thus drilling bore-holes for water can lead to overgrazing and, in consequence, exhaust pastures. The recent history of the Sahel provides a sad example.

Ecologically sound management of many arid and semiarid areas has yet to be achieved, and pastoralism in its mobile and extensive forms remains the most suitable form of economy for many arid areas of the world even as modernization has turned out to be very painful for the pastoralists themselves. What will happen to them in the future? I do not have an answer. Obviously many, if not everybody, will agree that the aim of anthropology is to understand the past and to explain the present, but in no way to predict the future.

I can only define the problem: how to reconcile mobile pastoralism with modern trends and necessities of life without too much loss not only for the pastoralists themselves but for mankind in general. I must also emphasize that pastoralism is not only a way of making a living; it is also a way of living, dear to those who practice it. As Webster Robbins, a Native American, remarked very aptly during the symposium: "You will not have economic development until you permit self-development." Planners come, planners go. If only they did not prefer to plan in their armchairs. External subsidies and support are provided and withdrawn, but the pastoralists remain in their harsh natural environment and have to pay for ill-devised projects. That is why, though without much hope, I desire that the opinions of the pastoralists themselves be taken into consideration by the policymakers and the planners, those bright guys on the Hill who too often think they have the right to interfere in the lives of others and plan their destiny for them.

REFERENCES

Allworth, Edward, ed. 1967. *Central Asia: A century of Russian rule*. New York: Columbia University Press.

Ashmarin, A. F. 1925. Kochevye puti, zimovye stoibishcha i letovki. *Sovetskaia Kirgiziia* 5–6:III–25.

Aziatskaiia Rossiia. 1914. Vol. I. St. Petersburg: Pereselencheskoe Upravlenie.

Bekmakhanova, N. E. 1980. *Formirovanie mnogonatzionalnogo naseleniia Kazakhstana i Sevenoi Kirgizii*. Moscow: Nauka.

Bennett, John W. 1985. Range culture and society in the North American West. *Folklore Annual*, 88–104.

Dakhshleger, G. F. 1965. *Sotzialno-ekonomicheskie preobrazovaniia v aule i derevne Kazakhstana* (1921–29 god). Alma-Ata: Nauka.

Demko, George J. 1969. *The Russian colonization of Kazakhstan, 1896–1916*. Bloomington: University of Indiana Press.

Desiat let Kazakstana. 1930. Alma-Ata: Izdanie Gosplana Kazakhskoi ASSR.

Downs, James F. 1964. *Animal husbandry in Navajo society and culture*. Berkeley: University of California Press.

Dvorkin, B. Ya., and E. P. Maslov. 1971. *Kazakhskaia SSR: Ekonomiko-geograficheskii ocherk*. Moscow: Prosveshchenie.

Eisenstadt, S. N. 1983. Development, modernization and dynamics of civilization. *Cultures et Développement* 15:2

Ekonomicheskoe razvitie Kazakhstana i kritika burzhuaznykh falsificatorov. 1978. Alma-Ata: Nauka.

Ellis, Florence H. 1974. *Navajo Indians I: An anthropological study of the Navajo Indians*. New York: Garland.

Ewers, John C. 1955. *The horse in the Blackfoot Indian culture*. Bureau of American Ethnology Bulletin 159. Washington, D.C.: Government Printing Office.

Fedorovich, B. A. 1973. Prirodnye usloviia aridnykh zon SSSR i puti razvitiia v nikh zhivotnovodstva. In *Ocherki po istorii khoziaistva narodov Srednei Azii i Kazakhstana* (JIE, vol. 98). Leningrad.

Galaty, John G., Dan Aronson, and Philip Carl Salzman, eds. 1981. *The future of pastoral peoples*. Ottawa: International Development Research Centre.

Graivoronsky, V. V. 1979. *Ot kochevogo obraza zhizni k osedlosti (na opyte MNR)*. Moscow: Nauka.

Jackson, W. A. Douglas. 1962. The Virgin and Idle Lands Program reappraised. *Annals of the Association of American Geographers* 52 (1):69–79.

Khazanov, A. M. 1981. Myths and paradoxes of nomadism. *Archives Européennes de Sociologie* 22:141–53.

———. 1984. *Nomads and the outside world*. Cambridge: Cambridge University Press.

Komarov, Boris. 1978. *The destruction of nature in the Soviet Union*. London: Pluto Press.

McCauley, Martin. 1970. Kazakhstan and the Virgin and Idle Lands Programme, 1953–64. MIZAN 12:100–11.

Maliki, Bonfiglioli (Angelo). 1986. The changing structures of livestock ownership among pastoralists in Niger. *Bulletin of the Institute for Development Anthropology* 4(1):3–5.

Masanov, N. E. 1984. *Problemy sozialno-ekonomicheskoi istorii Kazakhstana na rubezhe XVIII–XIX vekov.* Alma-Ata: Nauka.

Murdock, George Peter. 1968. The current status of the world's hunting and gathering peoples. In *Man the hunter*, ed. Richard B. Lee and Irven DeVore, 13–20. Chicago: Aldine.

Neishtadt, S. A. 1960. *Ekonomicheskoie razvitie Kazakhskoi SSR (period sotzializma i razvernutogo stroitelstva kommunizma).* Alma-Ata: Kazakhskoie Gosudarstvennoe Izdatelstvo.

Nurmukhamedov, S. B., V. K. Savosko, and P. B. Suleimenov. 1966. *Ocherki istorii sotzialisticheskogo stroitelstva v Kazakhstane (1933–1940).* Alma-Ata: Nauka.

Oliver, Symmes C. 1962a. *Ecology and cultural continuity as contributing factors in the social organization of the Plains Indians.* University of California Publications in American Archaeology and Ethnology, 48(1). Berkeley: University of California Press.

———. 1962b. The Plains Indians as herders. In *Paths to the symbolic self: Essays in honor of Walter Goldschmidt.* Special issue, *Anthropology* 18:35–43.

Pipes, Richard. 1964. *The formation of the Soviet Union: Communism and nationalism, 1917–1923.* Cambridge: Harvard University Press.

Reining, P., ed. 1978. *Handbook on desertification indicators.* Washington, D.C.: American Association for the Advancement of Science.

Salzman, Philip Carl. 1980. Preface. In *When nomads settle: Processes of sedentarization as adaptation and response,* ed. Philip Carl Salzman. New York: Praeger.

Underhill, Ruth M. 1956. *The Navajos.* Norman: University of Oklahoma Press.

Uteshev, A. S., and S. E. Semenov. 1967. *Klimat i vetrovaia eroziia pochv.* Alma-Ata: Kainar.

Vogt, Evon Z. 1961. Navaho. In *Perspectives in American Indian culture change,* ed. Edward H. Spicer. Chicago: University of Chicago Press.

Webb, Walter Prescott. 1931. *The Great Plains.* New York: Grosset and Dunlop.

Wedel, Waldo R. 1961. *Prehistoric man on the Great Plains.* Norman: University of Oklahoma Press.

Part Two

Traditional Use and Modern Alienation

The basic issues covered in this section are described in the Introduction. A few remarks may be added here. For the low-energy cultures of pastoral nomadic or hunter-gatherer peoples, the basic resource has to be land. In the four essays below, Russel Barsh, Gary Anders, Robson Silitshena, and Peter Iverson look at the transformation of fundamental land-use institutions in the Great Plains, Alaska, southern Africa, and Australia.

Institutionally, the change in land use goes from open-range hunting or herding of indigenous species to open-range herding of European breeds to fenced cattle feeding and plow agriculture. Ecologically, this transition means a reduction in the variety of flora and fauna and in the ability of stock to support itself on naturally produced resources such as native grasses. Monocultures tend to dominate, and often heavy soil erosion follows the introduction of the plow, as Robson Silitshena observes. As this movement across the "ecological transition" is taking place—and to facilitate it—property holding goes from a usufruct system, often highly sensitive to the need to use varying ecosystems in differentiated ways, to the European fee-simple system of free proprietorship that may make such usage difficult. Indigenous people whose lives have centered on animal consumption are likely to turn to some version of domesticated cattle raising and may be quite successful at it (for example, the Pine Ridge Lakotas were very successful with cattle up until the First World War, when they were forced to farm).

Variations obviously occur. Where a mixed agricultural/pastoral regimen is practiced—for example, by the tribes of the eastern plains such as the Omahas or the Pawnees of the nineteenth century or the southern African tribes described by Silitshena—the initial transition to European plow agriculture may not be difficult, but mechanized agriculture with its heavy machines may strain the indigenous system, particularly in the absence of institutions for extending credit to "tribal" peoples. Again, sedentarization in Alaska was much more a product of economic than of military force, and the struggle for Alaskan aboriginal land, mostly not valuable for farming, did not begin in earnest until its mineral wealth was discovered.

Present controversy focuses on land rights. The countries in which indigenous people are citizens, and have full rights before the law, obviously exhibit a different pattern from countries such as South Africa, where comparable rights do not exist. In countries such as Australia and the United States, land rights governed by law may bring a recrudescence of hope and even of certain aspects of indigenous culture, particularly on reserves that have had some success in defending their resources and institutions, as Peter Iverson shows in his essay.

The primary worldwide problem posed by this section is how to create an international system of land rights that permits indigenes to continue to use semiarid locales where traditional regimens have not been destroyed and permits experimentation with methods inspired by traditional systems in other areas. The essays by Barsh, Anders, Silitshena, and Iverson explain how complex these issues are.

Chapter Three

The Substitution of
Cattle for Bison on
the Great Plains

Russel L. Barsh

My father the sun, have pity on us!
—Arapaho buffalo-hide song

The sun dance formed the spiritual center of life for the Arapahos, in common with most of the peoples of the Great Plains who lived primarily by
keeping horses and hunting, and buffalo formed the spiritual center of the
sun dance. Dancers fixed their eyes on a buffalo skull atop the center pole, to
which they were tethered by buffalo-hide thongs, and the offering of flesh
was accompanied by a buffalo-hide song in which the ceremonial sacrifice of
human "hide" and the life-giving sacrifice of buffalo meat were symbolically
equated. Long after the buffalo economy had disappeared, however, the sun
dance survived to go underground in the 1890s and reemerge in the 1940s.
The buffalo were absent, but the sun dance message, describing the central-
ity of sun, water, and life, retained its meaning for most of the "pastoral" bison hunters of the Plains as government policy forced them to turn to farming and ranching. The process by which this turning was imposed is worth
exploring, as is its meaning for the environment.

Four centuries ago, 30 to 40 million American bison ranged from the
Rockies as far east as Pennsylvania and Ohio and included subspecies
adapted to both the grasslands and the woodland environment (Lott 1981;
McDonald 1981; fig. 3.1).[1] By 1875 there remained only two western grassland

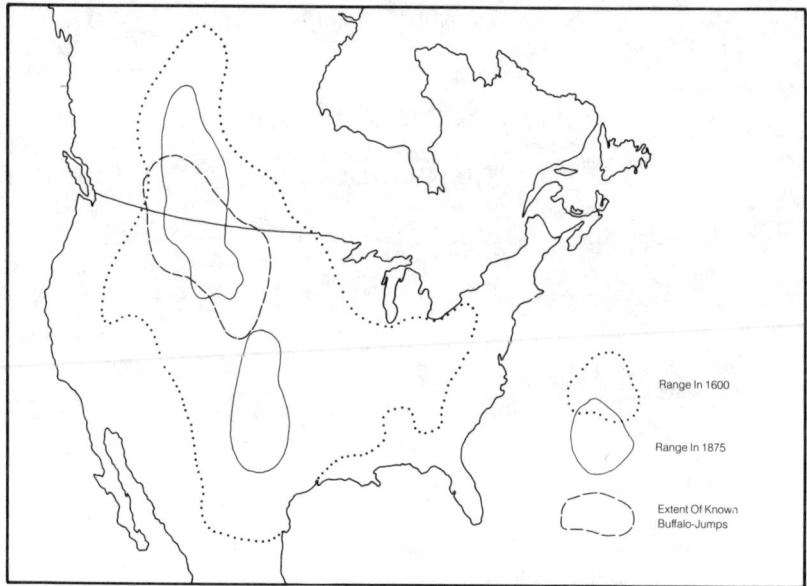

Figure 3.1 Distribution and paleo-Indian exploitation of American bison (*Bison bison*).
Data from Allen (1876) and Reeves (1983).

concentrations, the "northern" and "southern" herds, and by 1890 bison were practically extinct in the wild, replaced over most of their original range by domestic cattle. The substitution of cattle for bison had catastrophic effects on the aboriginal inhabitants of the Great Plains as well as significant implications for Plains ecology.

Current-day appreciation of the role bison played on the Great Plains is hampered by the necessity of studying small populations on preserves and reconstructing from inconsistent historical reports. Early writers such as Allen and Dodge characterized bison herds as highly organized and going on great annual north-south migrations, but this now appears unlikely on historical and ecological grounds. Unlike cattle, bison are relatively unsociable, aggregating randomly at wallows, rich pastures, and water. Herd movements are relatively short, opportunistic, and unpredictable. This strategy maximizes dispersal and thus reduces the risk of starvation in a habitat that, although it is a relatively stable climax plant community, experiences great local variability in microclimates and forage productivity.[2]

Male bison are promiscuous, remaining separate from the larger cow-calf herd except during a brief midsummer rutting period about a month after calving, and there is no evidence of extended cow-calf relationships. Dominance relationships appear only at times of scarce forage and then seem to favor the older animals. In the past, both young and old animals were continually culled by wolves, the bison's chief predator, but the only significant factor limiting bison numbers in the Great Plains probably was the availability of forage based on rainfall in spring and summer and temperature and snow depth in winter. Herds varied unpredictably from year to year, but total numbers probably expanded and contracted with the long-term drought cycles characteristic of the Great Plains (Hudson and Frank 1986; Bryson 1974). A relatively wet period ending about 1850 coincided with the rise and fall of mounted Indian bison hunting.

These characteristics of bison ecology determined much of the character of early Plains Indian cultures. Indigenous Americans were efficient bison harvesters for more than five thousand years before the introduction of Spanish horses, using natural buffalo jumps as well as corrals and rows of rock cairns arranged to funnel the easily stampeded animals (Reeves 1983). The southern and eastern tribes fired the prairies to achieve the same result (Grinnell 1892, 279–80; Allen 1876, 202–5). Although great numbers of bison could be taken quickly by these methods, hunters had to wait for the herds to come to them, resulting in considerable unpredictability and great waste after a successful hunt.

Though horses were not necessary for bison hunting, they made it possible to follow bison herds over a wider range and to exploit them year-round. Since horses were also an improvement over dogs for transporting meat, they permitted larger, more stable social groups to emerge without much increase in the available food supply or the total human population (Hanson 1986). At an average of no more than two horses per capita or, based on the 1865 Plains Indian population, a total of fewer than 150,000 domesticated horses in historical times, horse herding would have had negligible effects on forage availability (USDI 1865, 589–90). The number of wild horses may have been much larger, however. The ecology of feral horses is still not well understood, but like bison they appear to be eclectic foragers capable of exploiting a wide range of habitats and vegetation (Ganskopp and Vavra 1986; McInnis and Vavra 1987). This suggests that horses and bison could have coexisted with relatively little competition.

Indians intensified their harvest effort in early autumn when the animals were fattest and their hides in best condition, and they frequently took only the most tasty and nutritious parts, such as humps and tongues (Garretson 1938, 97; Dodge 1882, 291). This was not necessarily wasteful so long as it did not result in exceeding the herds' harvestable surplus. Polar bears also respond to the abundance of prey by hunting more often and eating more selectively, enlarging the scavenger niche at the same time (Stirling 1975). On the Great Plains, men and wolves were copredators, since more efficient human hunting produced more food for wolves as well. As long as the human population remained small relative to bison stocks, moreover, the ecosystem remained stable. In the early nineteenth century, based on contemporary estimates of 65,000 to 150,000 Plains Indians and at least 10 million bison, the ratio of human to bison biomass would have been only about 1:67,000.[3]

Indian hunters targeted the more numerous bison cows because their hides and meat were superior. If one assumes a total edible weight of 250 kilograms per bison cow, a nutritional value of 1,200 calories per kilogram,[4] and estimates of the minimum energy requirements for active humans (at least 2,800 calories per day), Plains Indians would have required 3 to 4 bison per capita yearly to survive. Even if they discarded all but the humps and tongues, reducing the utilized weight per animal to 50 kilograms or less, the minimum requirement would have been only about 15 to 20 bison per capita or, as Allen estimated by different means, a total of 1 million bison yearly. Although fur trading offered additional incentives for bison-hunting, it averaged only about 100,000 hides yearly in the 1850s, when it was still dominated by Indians (Allen 1876, 185–88). One can then think in terms of a total Indian harvest of a little over 1 million bison yearly, less than 10 percent of the stock on the Plains.

The "sustainable yield" of bison cannot readily be determined. The maximum proportion of an animal population that can be harvested safely depends on a number of biological parameters, such as age at sexual maturity, fertility, brood size, longevity of adults, and mortality from other causes, but a harvest rate of 50 percent of the unexploited population is generally taken as a first approximation for prolific small mammals (Eberhardt 1977). The rate should be somewhat lower in the case of large, longer-lived mammals such as bison.[5] In any case, a 10 percent exploitation rate for bison was probably very safe.[6] The natural rate of increase of contemporary bison herds has been calculated at roughly 18 percent (Van Vuren and Bray 1986), under conditions of minimal predation.

Early Indian farming cultures probably expanded and contracted in response to the long wet/dry cycles on the Plains (Wedel 1941). After the development of mounted hunting, however, farming remained restricted to the major river valleys, suggesting that bison hunters had gained sufficient economic and military power to defend their frontiers against more numerous, sedentary neighbors whose population was gradually declining from introduced diseases, as did the nomads of northern China a millennium earlier. In any case, Plains Indian social organization after the 1500s was, as Goldschmidt (1959, 209–16) and Oliver (1962) have argued, more characteristic of herders than of hunter-gatherers. Greater hunting mobility and productivity promoted the growth of much larger sociopolitical organizations, and Plains Indians' reliance on horses, a highly mobile and precarious form of wealth, encouraged raiding, defensive militarism, and male dominance.[7]

The forced transformation of this hunting culture began in earnest after 1870 as a result of economics and government policy. Although a limited trade in "buffalo robes" developed along the Mississippi River in the 1820s, it averaged no more than 100,000 bison yearly in the 1840s and 1850s (Allen 1876, 185–88). After the Civil War, however, the westward expansion of railroads created both a temporary demand for bison meat to feed construction crews and also improved transportation. Rising immigration and the 1873–75 depression produced a large surplus of itinerant labor, and the invention of a means of producing bison leather in 1871 increased the hides' usefulness as a cheap substitute for cow leather in industrial drive belts and military equipment. Animal bones were also in demand to make carbon for the new sugar-refining industry (Garretson 1938, 157; Dodge 1882, 293).

The great bison hunt of the 1870s was therefore more a product of temporary economic circumstances than of the intrinsic value of the hides. Some 5,000 men were briefly employed in the hide-hunting trade, and owing to the great abundance of both animals and hunters, a hide cost only $1 in 1874 (Allen 1876, 187). At about $100 for a rifle and 25 cents for each cartridge, each hunter required only a small "grubstake" and broke even at about 135 hides. Without incentives to conserve, hunters shot as many animals as they could, taking hides only from cows and young bulls and abandoning as many as five hides for each one shipped (Allen 1876, 177, 197; Dodge 1882, 293; Garretson 1938, 116).

Clearly the Great Plains were soon to be emptied of bison. The commercial hunt was more than double the estimated annual Indian harvest of 1 million animals, and by the autumn of 1873 "the vast plain, which only a short twelvemonth before teemed with animal life, was a dead, solitary, putrid desert" (Dodge 1882, 295). The Atchison, Topeka & Santa Fe Railroad alone shipped 251,443 hides, 1.6 million pounds of meat, and 2.7 million pounds of bones in 1873, and 7 million pounds of bones in 1874. The southern bison herd had disappeared by 1878 and the northern herd by 1883, representing a loss of probably 10 million animals (Andrist 1964, 182; Osgood 1957, 89; Allen 1876, 190).

Americans condoned the slaughter, which principally benefited western cattlemen. Bison, "whose hoofs and horns meant death," competed with domestic cattle for forage and sometimes stampeded them, though the competition was not always exact because, as I shall indicate, the two animals' forage habits differed (Henry 1930, 8; USC 1876, 1239). The ranch industry systematically also destroyed wolves and prairie dogs to make the range safer for cattle. Westerners viewed the elimination of bison as "a great step forward in the civilization of the Indian," moreover, because Indians frequently left their reservations to pursue roaming bison herds (USC 1876, 1239; Osgood 1957, 79). As one observer put it, "So long as there are millions of buffaloes in the West, so long Indians cannot be controlled even by the strong arm of the Government." Despite moral aversion to the use of starvation as a means of subjugating the tribes, Americans generally shared Texas congressman James Throckmorton's attitude that "the more buffaloes are exterminated the better it will be for our country" (USC 1876, 1239). "If they stand up against the progress of civilization and industry, they must be relentlessly crushed" (USDI 1872, 9). A proposal to restrict bison hunting to Indians passed the House in 1876 but died in the Senate.

Having tolerated the destruction of the Indians' own food supply, the government ironically found itself buying beef from cattlemen to feed them (USC 1876, 1238). In 1889 the United States Indian Service paid $768,000 for 31.5 million pounds of ration beef, or 485 pounds per capita, for Plains Indian reservations alone (USDI 1889, 496–512, 554–69), about 1 percent of total United States beef production for that year and as much as 10 percent of western production (computed from USBC 1975, series K195). In turn, many government beef contractors leased reservation land from the Indian Service at an advantage, and beef remained the principal ration item and a major part of the total cost of administering the reservations until the First World War (USC 1914, 1773; Dale 1960, 64). In 1900, 45,270 Indian households were receiving up to a maximum monthly allowance of 150 pounds of beef or bacon, 50 pounds of flour, 7 pounds of sugar, 4 pounds of coffee, and 3 pounds of beans (USDI 1900, 6–7).

Although the Grant administration's motto was "better feed the Indians than fight them" (USDI 1868, 20), Indian agents quickly seized on the ration system as a disciplinary tool. At Pine Ridge, V. T. McGillicuddy reasoned that since "'Poor Lo,' not unlike his white brother, is peculiarly sensitive in the gastric region," Indians could be induced to labor by reducing their rations (USDI 1882, 36). Although some of his colleagues condemned the "starving process" (USDI 1881, 80), rations were gradually reduced in the 1880s, and by 1900 it was Indian Service policy to adjust rations to need and to issue sugar and coffee only as a reward for Indian labor (USDI 1882, 60; 1900, 8).

As late as the 1870s, Indians on the northern plains were still able to obtain an estimated one-fourth of their subsistence from bison (USDI 1881, 80), but by the 1880s they were turning to hunting cattle to avoid starvation.[8] The Indian Service unwittingly encouraged this transition by issuing the beef ration on the hoof so that the older Indians could chase them down and shoot them, providing entertainment for visiting whites (USDI 1892, 454, 461; 1900, 8).

For Indians, the disaster was not only socioeconomic, but also nutritional. Indian agents early recognized that range-fed Texas steers were leaner and therefore more like bison than corn-fed beef (USDI 1877, 4–5), and modern studies confirm that bison, like other wild meats, is significantly lower in both total fat and saturated fat than feedlot beef (Crawford 1975, 464–69). Both feedlot and range beef were issued to Indians, and government rations contained both more saturated fat and more sugar than the aboriginal diet, the kind of diet change that has been linked to a threefold increase in the risk of heart disease among Inuit since 1950.[9] It probably took a similar toll among Plains Indians a century ago.

Government land policy at first favored the free-range cattle industry and only later encouraged fencing.[10] The American range cattle industry had its origins in the Southwest well before the Civil War.[11] Aided by liberal land grants from the Mexican and Texan republics, Texas cattle had increased to at least 3 million by 1860. Access to markets was a problem, however, although cattle were occasionally driven, at prohibitive costs, to Chicago and California. The Civil War changed all that by destroying much eastern farmland, raising prices and making Texas cattle one-fifth as expensive as eastern cattle in 1866. The Civil War wrought an estimated 50 percent decline in the capital value of southern farms (Lindert 1988, 54, 74), and the loss of manpower in the North was barely offset by an increase in the use of machines (Bogue 1963, 156, 165). After expanding by 39 percent in the 1850s, United States farm acreage grew by less than 1 percent during the 1860s, reflecting a loss of 36 million acres in the South, little change in the Northeast, and a gain of 31 million acres in "corn belt" states such as Illinois and Iowa (USBC 1975, series K17–81). Stock raising quickly expanded into the northern plains, attracted by the "leagues of free grass" newly liberated from bison herds and Indians, stimulated by the demand for beef from emigrant trains, the army, Indian Service, and mining settlements, and facilitated by the westward expansion of the railroads. Refrigerator cars, introduced in 1869, greatly improved the value of Texas stock fattened on northern ranges before they were shipped east through Chicago and Kansas City.

The introduction of refrigerated ships in 1875 opened new markets overseas, and exports of fresh beef jumped to 106 million pounds by 1881, to the dismay of British farmers already suffering since the 1860s from burgeoning exports of American wheat (USC 1888, 8–9; USBS 1889, 95; Trimble 1918). American beef exports continued to grow until 1900–1910, when shipments averaged more than 500,000 cattle yearly. Wheat, beef, and other agricultural products made up four-fifths of all American exports in the 1880s and declined in importance only after the end of the First World War, overtaken by American manufacturing and competition from Argentina and Australia (Trimble 1918).

Indian land played an important role in developing this bonanza: Indian treaties opened nine-tenths of the prairies to settlement in just four years following the Civil War (computed from Royce 1899). The government negotiated further reductions of many Montana and Dakota reservations in the 1870s and in 1887 began to sell "surplus" reservation lands under the General

Allotment Act. While cattlemen initially welcomed these changes, the spread of farmers put much former rangeland under plow and pressured the ranchers into seeking new land. By the 1880s many tribes were leasing large tracts to cattle companies, making their reservations islands of cheap, unfenced range in the midst of encroaching farmers (Burrill 1972; USC 1914, 1827; USDI 1889, 30). Indian Service beef contracts also contributed to the cattlemen's fortunes, composing about 10 percent of all sales of western cattle.

The availability of free range and rapid growth of eastern and European beef markets guaranteed cattlemen high prices and profits, increasingly attracting speculative—especially British—investment (Nimmo 1885, 44–46). Texas cattle were brought north in growing numbers to build larger herds, further inflating stock prices and reinforcing the illusion of limitless profitability. The number of range cattle in Wyoming alone tripled from 1877 to 1880. Rising prices also encouraged cattlemen to introduce costlier midwestern cattle breeds, increasing their financial risk. Although larger and more valuable, midwestern breeds were not as hardy or self-sufficient as Texas stock.

The most devastating result of this investment fever was overstocking the best range. By the 1880s, Colorado grass was already playing out, and herds were being redeployed north. The emphasis shifted from free grass to subdividing the commons and controlling access to scarce water. Barbed wire was patented in 1874, and its production doubled yearly from 1877 to 1880 as cattlemen hurriedly tried to fence the public domain in a race against the westward movement of the farm frontier.[12] Kansas farmers gradually forced the great north-south cattle trails westward, and by 1886 farmers had begun fencing the northern Plains.

Cattle companies had no alternative but to acquire title to land, at least for the production of winter feed, and no matter how cleverly they schemed to limit their purchases to a few strategically positioned parcels, expenses rose and profits fell. Since overstocking decreased the availability of summer forage, the animals were in poor condition, increasing winter losses. Fencing, in addition to the public lands it used, aggravated the problem by restricting the herds' winter movements, leading to even greater winter kills. During the severe winters of 1885–86 and even worse 1886–87, losses jumped from the "normal" 5–10 percent to 40–60 percent. Panicky cattlemen sold off their stock in 1887, prices crashed, there was a wave of bankruptcies, and foreign investment disappeared.

By 1895, western range stock had fallen by two-thirds, creating a vacuum quickly filled by sheep. Able to thrive on range already cropped by cattle, sheep peaked during 1895–1910, when they outnumbered cattle by a factor of six. For their part, cattlemen shifted from trying to control the range to cultivating winter forage crops. Hay acreage in Montana and Wyoming increased tenfold from 1890 to 1900, and western irrigation expanded rapidly, about 80 percent of it used to water hay. Despite breeding improvements, moreover, free-range cattle in 1883 were still only about two-thirds the body weight of grain-fed stock. After 1875 an increasing proportion of western cattle were finished on surplus midwestern corn, and by the 1890s the grain and cattle industries could be considered a single production system. Feeding reduced ranchers' risks and improved farm prices by raising the demand for grain.

One more boom-and-bust cycle came with World War I. Beef prices rose precipitously, triggering another round of herd expansion. When prices crashed in 1919, there was another cattle sell-off, and financing again grew scarce. In the hardest-hit southwestern prairies, landowners switched from cattle to dry wheat farming, encouraged by price supports and the introduction of tractors and mechanized plowing (Bonnifield 1979, 47–57). By the late 1920s Great Plains farmers were tilling an estimated 24 million acres of unstable or unprofitable old rangeland (Held and Clawson 1965, 81). The soil began to blow away, especially in the 1930s in the heartland of the old range cattle industry: Texas, Oklahoma, Colorado, and Kansas.

As the foundation of the range cattle industry had been free grass, that also was its downfall. Although customary rules helped subdivide the range among long-established cattlemen, no effective means of preventing new entry existed, and newcomers increased the cattle herds to the point where they exceeded the habitat's long-term carrying capacity. The homestead laws were no help, because they were designed for farmers rather than cattlemen. A 160-acre entry on the Great Plains could support, at best, five to fifteen cattle. Proposals to change the nation's land-tenure system to facilitate ownership of larger tracts and water access were generally regarded as monopolistic and collapsed in the face of a growing populism.

Although they were initially victims of the new cattle regime, Indians rapidly found ways to participate in it. Early Texas trail drives idled in the Indian Territory to take advantage of the region's rich grass. Some tribes insisted on expelling the cattlemen, while others, such as the Cherokee Nation and the Osages, charged tolls and later arranged favorable leases.[13] Indian agents recognized that ranching was more promising than farming on western reservations, not only for environmental reasons but because Indians themselves preferred it.[14] Between 1878 and 1881, the Indian Service supplied Indians with 13,264 head to start their own herds and was asking Congress to appropriate more funds for this purpose.

The continuation of the beef ration was an associated issue. "Under the nonsensical treaties at present in force," Pine Ridge agent McGillicuddy argued, "these people are guaranteed plenty of beef to eat, whether they work or not, so what earthly object has an Indian in going to the trouble and labour of raising beef?" (USDI 1883, 35). Other agents maintained that cutting back the beef issue would simply force Indians to kill their cattle, and he recommended distributing enough cattle to provide both food and breeding stock (USDI 1881, 52, 79; 1882, 66, 58). McGillicuddy's views prevailed and rations were gradually reduced, while cattle were issued only to the "most deserving and industrious Indians" in the hope that they would set an example for others.[15] Indians were forbidden to slaughter or sell any of these government-issue "I.D." cattle without the agent's permission, but this simply depressed their incentive to care for their stock, and restrictions on Indians' spending their own money made it difficult for them to acquire additional teams and cattle.

Although Indians were frequently discouraged by severe winter losses, tick fever, and white rustlers,[16] ranching had become the leading activity on most Plains reservations by the 1890s. Half the Crows already owned 5 to 35 head of cattle, and many Cheyenne River Sioux owned 50 to 100 head or more. Some Indian ranchers at Flathead and Cheyenne River began shipping directly to Chicago, and at Fort Belknap, Warm Springs, Crow Creek, and Pine Ridge they began filling their own tribes' ration-beef contracts (USDI 1889, 225; 1895, 174–78, 271–94). In Montana and Wyoming, Indian-owned cattle increased from 15,930 in 1880 to 179,635 by 1900 (USDI 1881, 292–301; 1900, 662–73).[17] Indian ranching gradually concentrated in a few families, some controlling thousands of cattle (USDI 1895, 271; USC 1914, 1907; Grinnell 1915).

Trespassing white cattlemen were a persistent problem, but the Indian ranchers' greatest enemy was the Indian Service itself, which began leasing the reservations' best forage and water to whites, fencing the leased range at the Indians' expense, and trusting the lessees to round up and brand intermixed Indian stock.[18] The Indian Service even included Indian allotments in grazing leases without obtaining the owners' consent and permitted sheepmen to operate on rangeland already fully stocked with Indians' cattle (USC 1914, 1925, 1961). On the Crow reservation, Indian-owned cattle declined from 12,000 to no more than 2,000, while lessees brought in 50,000 cattle and 45,000 sheep (USC 1914, 2115).

As the Indians' range shrank, the Indian Service inaugurated a program of "tribalization" of cattle herds that was supposed to improve management, stimulate demand for reservation-grown forage crops, and incidentally prevent Indians from eating their own herds (USC 1914, 1753–74, 1905–6). In fact, this compulsory pooling of stock eliminated individual initiative and placed Indians' hard-won new wealth in the hands of incompetents. On the Northern Cheyenne reservation, where serious ranching only began in 1906, Indians owned 460 individual brands and enjoyed an aggregate annual net income of more than $30, 000 by 1914 (USC 1914, 1752, 1766, 1868). After tribalization, their herd shrank to 4,200 head. On the Blackfeet reservation, individual holdings of more than 60,000 head fell to 2,000 (USC 1929, 12601, 12684, 12846). This tragedy was not Garrett Hardin's "tragedy of the commons," where common lands were overused; it was instead the tragedy of the destruction of tradition where traditional roles in the handling of common resources had been altered or destroyed.

Just as Indian ranchers were beginning to realize substantial profits in the decade preceding World War I, moreover, the Indian Service began to encourage farming in the belief that it would lead to greater employment than stock raising (Pennington 1978, 14–15). Instead of putting up hay for their stock, Indian allottees were directed to plant food crops (USC 1914, 1974; 1929, 12522–30, 12751). The Indian Service accelerated the leasing of the best reservation grazing land and pressured Indians to sell their allotments.[19] Although allotment divided Indian rangelands into 160-acre parcels, many Indian ranchers pooled allotments with their kinsmen or grazed the unallotted tribal range. In the 1910s the Indian Service actively discouraged ranching, however, because farming was more labor intensive (Barsh 1987). This program coincided, interestingly, with the earliest government efforts to impose production controls on non-Indian farmers (Hall 1980). The Indian Service also began taxing allottees for irrigation projects that in fact benefited white farmers and ranchers.[20] As late as 1942, only one-third of the irrigated land on Indian reservations was Indian operated (USDI 1943, 28).

Crowded onto diminishing portions of their own reservations, restricted from harvesting adequate winter forage, and taxed for irrigation ditches, Indian cattlemen were the greatest victims of the 1919 agricultural depression. Mechanization delivered them another blow in the 1920s (Dempsey 1978, 28). By 1925, whites owned an average of more than one tractor per farm in the Great Plains (USBC 1928, 96), but Indians were unable to obtain the necessary financing, and the equipment purchased for them by the Indian Service, often with their own money, was so inferior that some congressmen suggested a conspiracy with the manufacturers (USC 1914, 1959–60; 1929, 12323, 12551–64). Revolving Indian agricultural credit programs were introduced as part of the "Indian New Deal" in 1934, but though this increased the number of Plains Indian cattlemen roughly fivefold, to 3,502 by 1942, the total number of Indian cattle was still only 40 percent of what it had been in 1900 (USDI 1943, 26–27; 1900, 662–73; USC 1919, 748).

Although Indians continued to use most of their own grazing land, average productivity was only about one-fifth that of whites' land as recently as 1974, and livestock income amounts to barely one-fifteenth of total Plains Indian per capita income, compared with one-third in the 1930s (AIPRC 1976, 40–44). Some of this has been due to a postwar contraction in ranching and renewed "tribalization" of the Indian range, but most reflects the rapid growth of Indian employment in federal reservation social programs, which now account for one-third of all reservation Indian income nationwide (Barsh and Diaz-Knauf 1984).

Substituting cattle for buffalo has had environmental as well as human consequences.[21] In unmanaged grasslands, insects and nematodes are the most important consumers of vegetation, but grazing animals also have significant effects on the structure of the plant community and the composition of soils. Grazing reduces the proportion of taller grass species and removes standing dead material, resulting in less mulching and lowering the organic content of the soil (Wood and Blackburn 1984). On the other hand, plant growth initially increases in response to grazing, and herbivore nutrition is

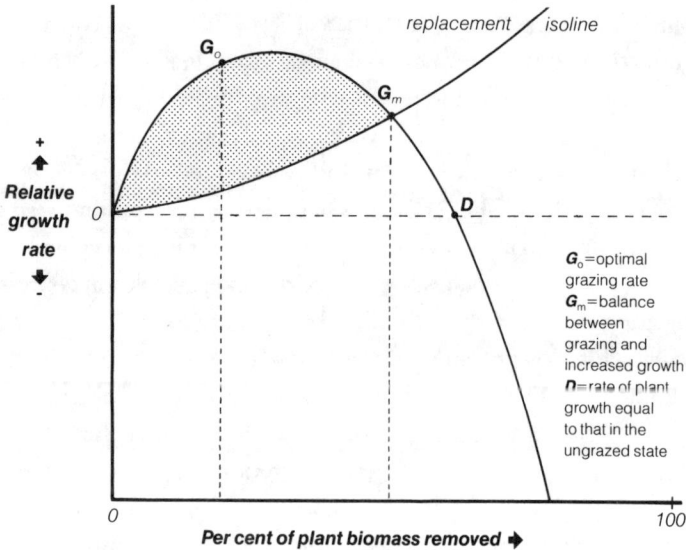

Figure 3.2 Hypothetical determination of optimal stocking rate. The O-G_o-G_m-D line represents the observed relation between the growth of forage plants and the intensity of grazing. The replacement isoline shows how much additional growth is needed at any given level of grazing simply to replace what has been eaten. After Dyer et al. (1979, 5).

therefore greatest at moderate rather than maximum grazing intensity. Wide dispersal and frequent movement of wild herbivores reduces grazing intensity, whereas the fencing and herding of domestic cattle can increase it to the point of reducing the productivity of forage plants (fig. 3.2).[22]

No discussion of bison ecology would be complete without considering the trophic relation between bison and prairie dogs (*Cynomys ludovicianus*), another herbivore that was conscientiously destroyed by early cattlemen. In the early stages of colonization, a new dog town removes standing dead matter and reduces the ratio of graminoids to forbs and shrubs in the plant community, but the protein content and digestibility of remaining graminoids increase (Garrett and Franklin 1988; Coppock et al. 1983). Present-day bison herds tend to spend most of their grazing time around young dog towns, where the forage is richest. Killing prairie dogs eliminated these islands of high-quality forage and thereby increased the geographical area that wild or domesticated herbivores would have to cover to obtain the same nutrition. Although the slaughter of bison preceded the efforts to control prairie dogs, these forage-quality effects probably influence the modern productivity of the Plains in relation to cattle as well as wildlife such as antelope and feral horses (McInnis and Vavra 1987).

At its peak, the free-range cattle industry filled only part of the habitat room left by bison. Americans owned some 22.5 million beef cattle in 1883, of which about 5.8 million were in the plains states (McCarty 1883, 249). At the average slaughter weight of 350 kilograms (Dale 1960, 151) for range cattle of this period, this number of cattle represents a Plains ruminant biomass of about 2 billion kilograms, compared with a bison biomass, at 500 kilograms per animal, of 5 billion kilograms, if 10 million of the estimated 30–40 million total bison occupied the Great Plains area (Rutberg 1984). The loss of herbivore production is even more striking if cattlemen's extermination of about 98 percent of the prairie dog population, another 5 billion kilograms biomass, is included (Coppock 1983).

There may be several behavioral and physiological reasons why cattle were less successful than bison on the Plains. Bison digest low-quality, nitrogen-poor forage better than cattle do, which means they can utilize more of the habitat, are affected less by overstocking, and can survive winters better (Hudson and Frank 1987; Keith et al. 1981; Peden et al. 1974). Bison also move more frequently than cattle and tend to be less selective in their choice of forage (Reynolds, Hansen, and Peden 1978; Ralphs, Kothmann, and Merrill 1986). They are hardier than cattle, are adapted behaviorally to surviving winter storms, and reputedly can detect water on the dry plains (Garretson 1938, 46). For European settlers, the chief advantages of cattle were manageability and familiarity rather than hardiness. That is, whites found it easier to replace bison than to discover a way of domesticating them. Indeed, cattlemen argued that bison were inherently wild: "They take up as much room and consume as much provender as cattle and horses," a Texas congressman argued in defense of the slaughter, but cannot be "domesticate[d] so as to make them useful" (USC 1876, 1239).[23]

Since cattle are more vulnerable than bison to lack of forage and severe winters, stocking range cattle up to the same level of biomass would simply have resulted in more winterkills and greater risk. Only through intense cultivation of feed grains and forage crops have Europeans been able to restore the aboriginal meat productivity of the Plains system, and there are considerable environmental costs. Frequent deep plowing has placed a significant proportion of the Great Plains at risk from wind and flood erosion, and maintaining crop productivity requires an increasing load of fertilizers, pesticides, and herbicides. On the whole, then, the restoration of the Plains ecosystem's productivity should be viewed as temporary.

Taking official stock of the range cattle industry a century ago, Nimmo argued that it had boosted exports, provided cheap meat for "the laboring people of this country," and "served as an indirect but effectual instrumentality in keeping the Indians upon their reservations, by expelling their game—chiefly the buffalo—from the ranges" (Nimmo 1885, 53). In retrospect, the most important economic contribution of the range cattle industry may have been the last one. Despite the heroic proportions the cattleman has assumed in American folklore, the industry itself contributed significantly to the gross national product for only a decade. In 1900 it directly employed a mere 48,015 men in the Great Plains states, or only 1.8 percent of all employment in the region, after eliminating the livelihood of more than 65,000 Indians (USBC 1904, 94–105). The real payoff of ranching was supporting the development of mining and other extractive industries in the Denver, Butte, and Deadwood areas, which had become the region's population centers by 1910.

The ecological significance of the bison/cattle transition was considerable. Overstocking the range, followed by intensive forage and grain farming, set wind and water erosion to work on what had been a relatively stable habitat, resulting in damage to soils over about one-third of the Great Plains by the 1930s (Held and Clawson 1965, 83). The region was of marginal agricultural significance in the first years of this century, with less than one-fifth the grain and forage production per square mile of the adjacent "corn belt" region, one-tenth the population density, and only one-third the cattle production, and so it has remained (USBC 1903, plates 13, 147, 153, 162). Indeed, it was already *losing* population by 1900, except around the mining centers. The range cattle boom served as one small outlet for the country's great oversupply of labor during the transition years from an agrarian to an industrial economy, but at the sacrifice of an enormously productive grassland ecosystem.

This boom also occurred at the expense of a civilization. Over the course of less than four centuries (1500–1900), Plains Indians were able to shift from hunting wild cattle, to herding horses in order to hunt wild cattle, to herding horses and hunting domestic cattle, to herding domestic cattle, but in the end they were worse off than when they began and were left with less than one-tenth of their land. The government's plans to make the Plains Indians self-supporting farmers in a single generation with a minimum of financing belied its naïveté regarding the entire Plains environment—particularly as to the equilibrium human population that could be maintained on renewable resources alone. The most remarkable thing about it may well be that the Indians survived the period 1880–1940 as well as they did, without more severe starvation and with a small increase in numbers, while whites began to abandon the rangeland within scarcely a generation of the first great bison hunt in 1873.

NOTES

1. Primary data on bison ecology in the first section of this chapter come from Roe (1970), Lott (1981), Grinnell (1892), Allen (1876), Dodge (1882), Hanson (1984, 1986), Lott and Minta (1983), and Reinhardt (1985). References in the text use the abbreviation USDI for United States Department of the Interior, USC for United States Congress, USBS for United States Bureau of Statistics, USBC for United States Bureau of the Census, and AIPRC for the American Indian Policy Review Commission. Estimates of nineteenth-century bison herd size are based on modern studies of the carrying capacity, or maximum stocking rate, of Plains habitats. Herd size could also be estimated from nineteenth-century harvest rates if they were known more reliably. See, for example, Thompson (1966).

2. Although Great Plains bison are chiefly grass eaters, Alaskan bison are more browsers (shrub eaters) than grazers (Waggoner and Hinkes 1986), probably much like the long-extinct bison of the eastern woodlands (McDonald 1981), illustrating the great dietary adaptability of this species.

3. Firearms naturally increased Indian hunting efficiency (less work per kill), but it does not follow that Indians increased their harvests. While an increase in the beaver hunt, for example, may have occurred as a result of access to European markets for furs, there was no comparably large new market for buffalo robes, hence no incentive to kill more buffalo. It is just as logical to argue that guns make it possible for Indians to spend more time doing other things—including hunting for other animals to sell their furs.

4. Only about half the live weight of domestic cattle (*Bos taurus*) is edible musculature, and since bison and cattle have similar musculature (Berg and Butterfield 1975, 37), it is reasonable to apply the same ratio to the average live weight (Rutberg 1984) of bison cows. As for calories, large free-living herbivores such as bison accu-

mulate less stored fat than feedlot cattle. Crawford (1975, 464) gives 120–30 calories/100 grams fresh meat for free-living herbivores, including range beef, and 230 calories/100 grams for "intensive fat beef."

5. Harvest rates based on average size of a population should be figured conservatively, because all animal populations are subject to unpredictable fluctuations.

6. A different conclusion would have to be drawn, of course, if the total bison population was not 30–40 million, as was proposed above, but fewer than 10 million. Recent studies of bison forage efficiency (e.g., Hudson and Frank 1987) support the higher estimates of nineteenth-century bison populations, however.

7. The spread of mounted hunting coincided with significant declines among agricultural villages exposed to European diseases. Dispersal may have promoted the survival of Plains peoples by reducing the risk of transmission of disease, although frequent visiting among camps and annual tribal gatherings ensured that no one was completely isolated.

8. This is confirmed by Berthrong (1971); Osgood (1957, 141–43); Dale (1960, 130); USDI (1889, 235); and USC (1914, 1754, 1825, 1844).

9. See Schaefer et al. (1980) and Schaefer and Steckle (1975). Although Indian health statistics from the 1880s are dominated by chronic upper respiratory and gastroenteric infections, which are indicators of malnutrition and poor sanitation (Barsh, n.d.), the greatest official attention was given to tuberculosis, which achieved epidemic proportions with mortality as high as 25 per 1,000 of Indian population. The virulence of this disease may have been due in part to the novelty of Indians' exposure to the pathogen (Buikstra 1981; Grzybowski, Stylo, and Dworkin 1976), but it was certainly aggravated by environmental factors such as malnutrition.

10. Although the United States industry is described here, the Canadian experience was essentially the same in all but one respect: Alberta cattlemen were able to obtain large leaseholds on the public domain, somewhat mitigating the "commons" problem that resulted in such severe overstocking of American rangeland. See, generally, Breen (1983).

11. Much of the analysis of the free-range cattle industry here is based on Dale (1960) and Osgood (1957). To cite every instance of my indebtedness to them would unnecessarily clutter the text.

12. See Nimmo (1885, 43–44); Andrist (1964, 331); Osgood (1957, 184, 190–93, 203–4); Dale (1957, 93–94). Naturally, barbed wire also helped the larger ranching companies fence out smaller, independent ranchers.

13. For details, see Berthrong (1971); Burrill (1972); Dale (1960, 124–28, 136–37); Nimmo (1885, 15); USDI (1882, 56, 83, 136; 1883, 30, 135).

14. On the environmental conditions favoring reservation ranching, see USDI

(1881, 52, 79, 129, 144; 1882, 56, 83, 136; 1883, 30, 135). On Indians' preference for ranching, see USDI (1876, 112; 1881, xxxviii, lviii, 33; 1882, 99; 1892, 232, 279, 295).

15. For these policies, see USDI (1883, 60; 1884, 37; 1892, 90–91; 1895, 167, 283–94); USC (1914, 1846–47, 1905–6).

16. For background on these problems, see USDI (1881, 52, 58, 173; 1882, 83; 1883, 40; 1884, 44, 100; 1900, 379–80); USC (1914, p. 1953).

17. Over the same period, the Indian Service waged a campaign to round up and destroy Indian ponies and mixed-breed horses, which it considered a waste of good range (USDI 1895, 271; 1914, 8). Indians still preferred to raise horses, however, and Indian-owned horses increased, albeit more slowly, from 35,352 to 95,617; USDI (1881, 23; 1884, 183; 1895, 221); USC (1914, 1775).

18. For discussion of these points, see USDI (1882, 14, 58; 1883, 61, 113, 126; 1884, 183; 1889, 254); USC (1914, 1821–48, 1875, 1884, 1908, 1955).

19. See USC (1930; 1929, 12314–15, 12344–46, 12357–58, 12847).

20. See USC (1914, 1992–94; 1929, 12556, 12561; 1920, 1111).

21. The basic resources on which this section is based are Dyer et al. (1979) and Coppock et al. (1983).

22. As grazing removes more and more of plants' structure, plant growth first increases, then decreases relative to the plants' ungrazed condition. At G_o, the *net* increase (*shaded area*) is greatest; this can be considered the *optimal* grazing rate. At G_m, grazing and increased growth just balance; this can be considered the *maximum* grazing rate. Past G_m, plant biomass is declining—the habitat is overstocked.

23. In addition to experiments with cross-breeding cattle and bison, many western ranchers today keep a few domesticated bison as a novelty. Other supposedly "wild" herbivores have been domesticated in relatively recent times, for example, reindeer (from caribou), and it should be borne in mind that cattle themselves, like all domestic ungulates, were derived from wild species two to five thousand years ago.

REFERENCES

Allen, J. A. 1876. *The American bisons, living and extinct.* Memoir, 4. Cambridge: Harvard Museum of Comparative Zoology.

American Indian Policy Review Commission (AIPRC). 1976. *Task Force Seven: Reservation and resource development and protection.* Washington, D.C.: Government Printing Office.

Andrist, Ralph K. 1964. *The long death: The last days of the Plains Indian.* New York: Macmillan.

Barsh, Russel L. 1987. Plains Indian agrarianism and class conflict. *Great Plains Quarterly* 7(2): 83–90.

———. n.d. Ecocide, nutrition, and the "vanishing Indian." In *State violence and ethnicity*, ed. P. van den Berghe. In press.

Barsh, Russel L., and Katherine Diaz-Knauf. 1984. The structure of federal aid for Indian programs in the decade of prosperity, 1970–1980. *American Indian Quarterly* 8(1): 1–35.

Berg, R. T., and R. M. Butterfield. 1975. Growth of meat animals. In *Meat*, ed. D. J. A. Cole and R. A. Lawrie. London: Butterworths.

Berthrong, Donald J. 1971. Cattlemen on the Cheyenne-Arapaho reservation, 1883–1885. *Arizona and the West* 13(1): 5–32.

Bogue, Allan G. 1963. *From prairie to corn belt: Farming on the Illinois and Iowa prairie in the nineteenth century*. Chicago: University of Chicago Press.

Bonnifield, Paul. 1979. *The dust bowl: Men, dirt, and depression*. Albuquerque: University of New Mexico Press.

Breen, David H. 1983. *The Canadian prairie west and the ranching frontier, 1874–1924*. Toronto: University of Toronto Press.

Bryson, Reid A. 1974. A perspective on climatic change. *Science* 184(4138): 753–60.

Buikstra, Jane E., ed. 1981. *Prehistoric tuberculosis in the Americas*. Evanston, Ill.: Northwestern University Archeological Program.

Burrill, Robert M. 1972. The establishment of ranching on the Osage Indian reservation. *Geographical Review* 62(4): 524–43.

Coppock, D. L., J. K. Detling, J. E. Ellis, and M. I. Dyer. 1983. Plant-herbivore interactions in a North American mixed-grass prairie. *Oecologia* (Berlin) 56(1): 1–15.

Crawford, M. A. 1975. Meat as a source of lipids. In *Meat*, ed. D. J. A. Cole and R. A. Lawrie. London: Butterworths.

Curtis, N. 1907. *The Indian's book*. New York: Harper.

Dale, Edward Everett. 1960. *The range cattle industry: Ranching on the Great Plains from 1865 to 1925*. Norman: University of Oklahoma Press.

Dempsey, Hugh A. 1978. One hundred years of Treaty Seven. In *One century later: Western Canadian reserve Indians since Treaty 7*, ed. Ian A. L. Getty and Donald B. Smith. Vancouver: University of British Columbia Press.

Dodge, Richard Irving. 1882. *Our wild Indians: Thirty-three years' personal experience among the red men of the Great West*. Hartford, Conn.: Worthington.

Dyer, M. I., J. K. Detling, D. C. Coleman, and D. W. Hilbert. 1979. The role of herbivores in grasslands. In *Grasses and grasslands: Systematics and ecology*, ed. James R. Estes, Ronald J. Tyrl, and Jere N. Brunken. Norman: University of Oklahoma Press.

Eberhardt, E. E. 1977. Optimal policies for conservation of large mammals, with special reference to marine ecosystems. *Environmental Conservation* 4(3): 205–12.

Ganskopp, David, and Martin Vavra. 1986. Habitat use by feral horses in the northern sagebrush steppe. *Journal of Range Management* 39(3): 207–12.

Garretson, Martin S. 1938. *The American bison.* New York: New York Zoological Society.

Garrett, Monte G., and William L. Franklin. 1988. Behavioral ecology of dispersal in the black-tailed prairie dog. *Journal of Mammalogy* 69(2): 236–50.

Goldschmidt, Walter. 1959. *Man's way: A preface to the understanding of human society.* New York: Holt, Rinehart and Winston.

Grinnell, George Bird. 1892. The last of the buffalo. *Scribner's Magazine* 12(3): 267–86.

———. 1915. *The Indians of to-day.* New York: Duffield.

Grzybowski, S., K. Styblo, and E. Dworkin. 1976. Tuberculosis in Eskimos. *Tuberc* 57(4, suppl.): S1–S58.

Hall, Tom G. 1980. Government controls: How to understand the experience of World War I. In *Farmers, bureaucrats, and middlemen: Historical perspectives on American agriculture,* ed. Trudy H. Peterson, 279–95. Washington, D.C.: Howard University Press.

Hanson, Jeffrey R. 1984. Bison ecology in the northern plains and a reconstruction of bison patterns for the North Dakota region. *Plains Anthropologist* 29(104): 93–113.

———. 1986. Adjustment and adaptation on the northern plains: The case of equestrianism among the Hidatsa. *Plains Anthropologist* 31(112): 93–107.

Held, R. Burnell, and Marion Clawson. 1965. *Soil conservation in perspective.* Baltimore: Resources for the Future by Johns Hopkins University Press.

Henry, Stuart. 1930. *Conquering our great American Plains: A historical development.* New York: E. P. Dutton

Hudson, R. J., and S. Frank. 1987. Foraging ecology of bison in aspen boreal habitats. *Journal of Range Management* 40(1): 71–75.

Keith, E. O., J. E. Ellis, R. W. Phillips, M. I. Dyer, and G. M. Ward. 1981. Some aspects of urea metabolism in North American bison. *Acta Theriologica* 26(14): 257–68.

Lindert, Peter H. 1988. Long-run trends in American farmland values. *Agricultural History* 62(3): 45–85.

Lott, D. E. 1981. Sexual behavior and intersexual strategies in American bison. *Zeitschrift für Tierpsychologie* 56(2): 97–114.

Lott, D. E., and S. C. Minta. 1983. Random individual association and social group instability in American bison (*Bison bison*). *Zeitschrift für Tierpsychologie* 61(2): 153–72.

McCarty, L. P., ed. 1883. *McCarty's annual statistician, 1883.* San Francisco.

McDonald, Jerry N. 1981. *North American bison: Their classification and evolution.* Berkeley: University of California Press.

McInnis, Michael I., and Martin Vavra. 1987. Dietary relationships among feral horses, cattle, and pronghorn in southeastern Oregon. *Journal of Range Management* 40(1): 60–66.

Nimmo, Joseph, Jr. 1885. *Report in regard to the range and ranch cattle business of the United States.* Wasington, D.C.

Oliver, Symmes C. 1962. The Plains Indians as herders. *Anthropology* UCLA 8:35–43.

Osgood, Ernest. 1957. *The day of the cattleman.* Chicago: University of Chicago Press.

Peden, D. G., G. M. Van Dyne, R. W. Rice, and R. M. Hansen. 1974. The trophic ecology of *Bison bison* L. on shortgrass plains. *Journal of Applied Ecology* 11(2): 489–97.

Pennington, William D. 1978. Government agricultural policy on the Kiowa reservation, 1869–1901. *Indian Historian* 11(1): 11–16.

Ralphs, M. H., M. M. Kothmann, and L. B. Merrill. 1986. Cattle and sheep diets under short-duration grazing. *Journal of Range Management* 39(3): 217–23.

Reeves, B. O. K. 1983. Six millenniums of buffalo kills. *Scientific American* 249(4): 120–35.

Reinhardt, V. 1985. Quantitative analysis of wallowing in a confined bison herd. *Acta Theriologica* 30(7): 149–56.

Reynolds, H. W., R. M. Hansen, and D. G. Peden. 1978. Diets of the Slave River lowland bison herd, Northwest Territories, Canada. *Journal of Wildlife Management* 42(3): 581–90.

Roe, Frank Gilbert. 1970. *The North American buffalo: A critical study of the species in its wild state.* 2d ed. Toronto: University of Toronto Press.

Royce, C. C. 1899. Indian land cessions in the United States. *Eighteenth Annual Report of the Bureau of American Ethnology, 1896–97.* Part 2. Washington, D.C.: Government Printing Office.

Rutberg, A. T. 1984. Birth synchrony in American bison (*Bison bison*): Response to predation or season? *Journal of Mammalogy* 65(3): 418–23.

———. 1986. Dominance and its fitness consequences in American bison cows. *Behaviour* 96(1–2): 62–91.

Schaefer, O., and J. Steckle. 1975. *Dietary habits and nutritional base of native populations of the Northwest Territories.* Yellowknife: Science Advisory Board of the Northwest Territories.

Schaefer, O., J. F. W. Timmermans, R. D. P. Eaton, and A. R. Matthews. 1980. General nutritional health in two Eskimo populations at different stages of acculturation. *Canadian Journal of Public Health* 71(6): 397–405.

Stirling, I. 1975. The caloric value of white ringed seals (*Phoca hispida*) in relation to polar bear (*Ursus maritimus*) ecology and hunting behaviour. *Canadian Journal of Zoology* 53:1021–27.

Thompson, H. Paul. 1966. A technique using anthropological and biological data. *Current Anthropology* 7(4): 417–24.

Trimble, W. 1918. Historical aspects of the surplus food production of the United States, 1862–1902. *Annual Report of the American Historical Association* 1:223–39.

United States Bureau of Census (USBC). 1903. *Statistical atlas: Twelfth census of the United States.* Washington, D.C.: Government Printing Office.

———. 1904. *Special reports: Occupations at the twelfth census.* Washington, D.C.: Government Printing Office.

———. 1915. *Indian population in the United States and Alaska, 1910.* Washington, D.C.: Government Printing Office.

———. 1928. *United States Census of Agriculture, 1925: Summary statistics, by states.* Washington, D.C.: Government Printing Office.

———. 1975. *Historical statistics of the United States, colonial times to 1970.* Washington, D.C.: Government Printing Office.

United States Bureau of Statistics (USBS). 1889. *Statistical abstract of the United States, 1886.* Washington, D.C.: Government Printing Office.

United States Congress (USC). 1876. *Congressional Record* 4:1237–41.

———. 1888. *Cattle and dairy farming.* Part 1. House Executive Document 51, 49th Cong., 1st sess. Washington, D.C.: Government Printing Office.

———. 1914. *Hearings, joint commission to investigate Indian affairs.* Parts 14 (Tongue River reservation) and 15 (Crow Indian reservation). Washington, D.C.: Government Printing Office.

———. 1919. *Hearings, Indians in the United States.* Vol. 1. House Committee on Indian Affairs. Washington, D.C.: Government Printing Office.

———. 1920. *Congressional Record* 59:IIII.

———. 1929. *Hearings, survey of conditions of the Indians of the United States.* Part 23 (Montana). Senate Committee on Indian Affairs. Washington, D.C.: Government Printing Office.

―――. 1930. *Hearing, Rosebud Indians*. Senate Committee on Indian Affairs. Washington, D.C.: Government Printing Office.

―――. 1934. *Hearings, to grant Indians the freedom to organize*. Part 2. Senate Committee on Indian Affairs. Washington, D.C.: Government Printing Office.

United States Department of the Interior (USDI). 1865, 1868, 1872, 1876, 1877, 1881, 1882, 1883, 1884, 1889, 1892, 1895, 1897, 1900, 1914. *Annual report of the commissioner of Indian affairs*. Washington, D.C.

―――. 1943. *Statistical supplement to the annual report of the commissioner of Indian affairs, 1943*. Washington, D.C.: Government Printing Office.

Van Vuren, Dirk, and Martin P. Bray. 1986. Population dynamics of bison in the Henry Mountains, Utah. *Journal of Mammalogy* 67(3). 503–11.

Waggoner, V., and M. Hinkes. 1986. Summer and fall browse utilization by an Alaskan bison herd. *Journal of Wildlife Management* 50(2): 322–24.

Wedel, Waldo R. 1941. Environment and native subsistence economies in the Central and Great Plains. *Smithsonian Miscellaneous Collections* 101(3): 1–29.

Wood, M. K., and W. H. Blackburn. 1984. Vegetation and soil responses to cattle grazing systems in the Texas rolling plains. *Journal of Range Management* 37(4): 303–8.

Chapter Four

The Alaska Native
Experience with the Alaska Native
Claims Settlement Act

Gary C. Anders

Future land administration of the Alaska Native, in light of the passage of the Alaska Native Claims Settlement Act of 1971, may well parallel the predicament of the Indians following enactment of the General Allotment Act (Dawes Act) of 1887. Both pieces of legislation arose during a period of economic fluctuation and uncertainty, and were promulgated at a point in history when Indian-White contact was at a precarious state with regard to social adaptation. In addition, integrating the Natives into the dominant white culture was an expressed goal of each settlement.—Fuller (1976)

The fundamental conflict between indigenous and industrial societies over the use of land and natural resources touches numerous tribal and semitribal groups on practically every continent, as several essays in this volume document. The purpose of this chapter will be to compare the Alaska Native Claims Settlement Act of 1971 (ANCSA) with the General Allotment Act of 1887 (Dawes Act), looking at both as efforts to impose a private property system, capitalistic in form, on collectively oriented subsistence practitioners, in Alaska and in the Great Plains. This essay focuses on the Alaska Native experience with an imposed corporate regime and is complemented by Barsh's examination of the effects of buffalo destruction, sedentarization, "reserve imprisonment," and allotment on the Great Plains native cattle industry.

Given the space limitations and the general familiarity of most readers with the allotment system, my principal focus will be upon ANCSA. This approach will allow a closer study of a contemporary public policy.[1] However, for comparison, I include some reminders of how the General Allotment Act worked in the Great Plains.

The history of allotting land to Indians goes back to the Continental Congress, but not until Massachusetts Senator Henry M. Dawes and various philanthropic groups such as the Indian Rights Association began to advocate was allotment taken seriously as a policy proposal. The General Allotment Act signed into law on February 8, 1887, contained five basic provisions:

1. Indian reservations would be divided, with each tribal member receiving a grant of land consisting of 160 acres for family heads. Unmarried Indians over the age of eighteen and juveniles would receive lesser grants of 80 and 40 acres, respectively.
2. Allottees would receive fee-simple title to their holdings, but the lands were to be held in trust by the government for twenty-five years, during which time the land could not be alienated.
3. Indians would be given four years to make their selections, then selections would be made for them by representatives of the federal government.
4. United States citizenship would be conferred upon any Indian who maintained his allotment and adopted a "civilized" life-style.
5. Unallotted tracts of land would be declared surplus and sold by the government.

Allotment had a particularly harsh effect in the Great Plains, where the so-called Indian wars had just been completed. The Dawes Act, as soon as it was implemented, forced members of clans or traditional governing groups to live apart and created massive social fragmentation. It turned treaty-protected territory over which the tribes had at least limited sovereignty into property that could be alienated to white speculators through the manipulation of the Indian agency's trust responsibilities, and it compelled the Plains tribes to adopt a pattern of resource management inappropriate to the area. In response, many Plains tribesmen turned to cattle herding as the economic activity closest to the old buffalo hunt and most compatible with their former way of life. On reservation after reservation, government im-

position of 160-acre allotments forced Plains Indians to change their traditional pastoral nomadic ways. Moreover, allotment divided the land into
units that did not permit efficient grazing and ultimately led to the land's being taken over by white farmers and ranchers. Consequently Indian herds
confined to small "farms" also had problems with limited water resources
and overgrazing and so have not significantly benefited most Natives, who
were impoverished by the breakup of tribal ownership. Enlarging upon key
elements of a comparative framework for assessing federal/native settlement
models, the following discussion focuses on four important dimensions of
allotment and ANCSA:

1. Recipients of the settlement (individuals vs. corporate entities).
2. The permanence of potential transfers of ownership or development
   rights.
3. The extent of external imposition and internal ratification of the settlement.
4. Social and economic effects on the groups involved in the settlement.

SETTLEMENT RECIPIENTS

The Alaska Native Claims Settlement Act appears to assign land to groups
rather than to individuals as the Dawes Act did, but the appearance is decep-
tive. Although there are complex facets of ANCSA that provide allotments to
individuals,[2] the main thrust of the settlement is a combination of cash and
land awarded to state-chartered corporations. In return for extinguishing all
aboriginal claims, ANCSA provided a cash settlement of $962.5 million and
approximately 44 million acres of land. The act designated twelve Native regions in the state, each region reflecting common culture and language
whenever possible (map 4.1). Lands chosen in each of these regions were to
be managed by a regional corporation, with supplementary village corporations established in local communities with at least twenty-five residents.

The twelve regional corporations and a later thirteenth (established for
nonresident Alaska Natives) were incorporated as for-profit enterprises as
mandated by the terms of ANCSA. Also, almost all the village corporations
chose to become for-profit entities so that they could make cash distributions to their shareholders. Of the more than two hundred villages that
qualified for ANCSA benefits, only seven chose not to participate in the settle-

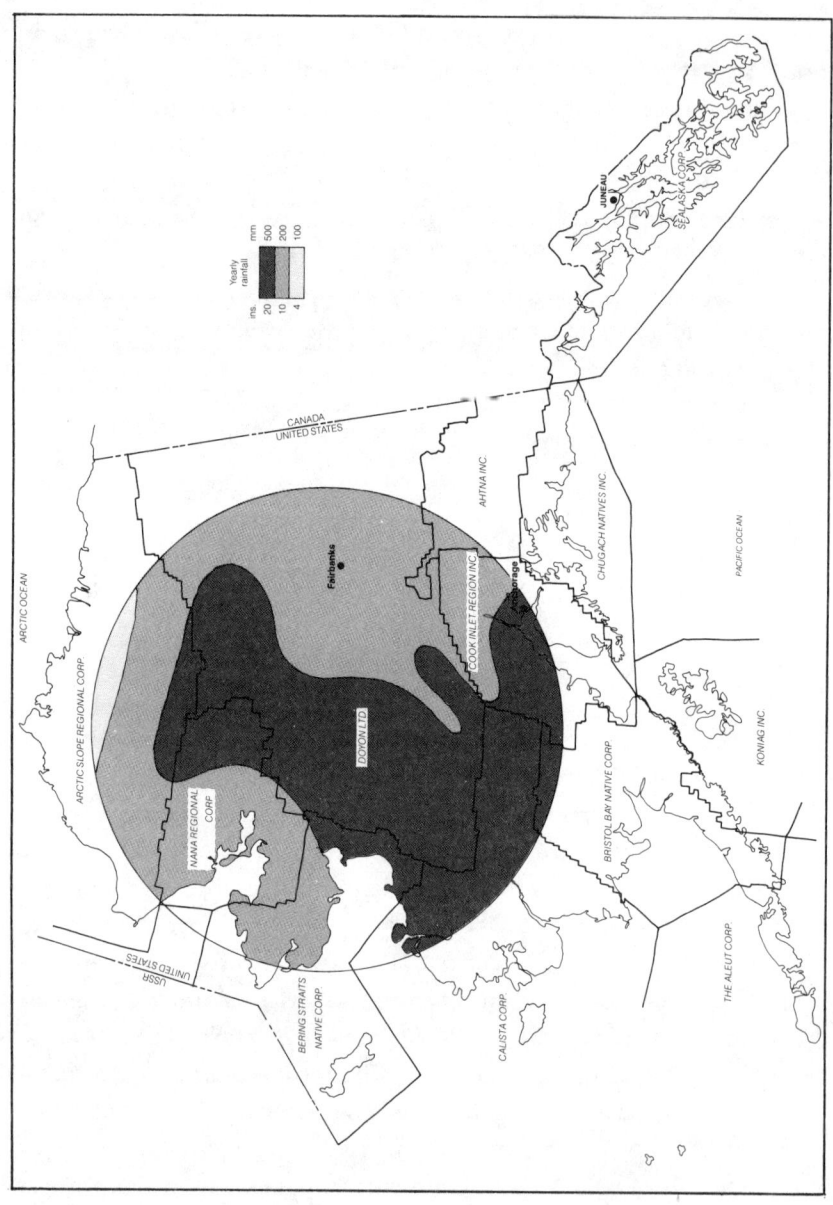

Map 4.1 Alaska Native corporation regions.

ment, instead selecting former reserve lands that were transferred to a traditional government or Indian reorganization council by the Indian Reorganization Act (IRA).[3] These seven villages also received relatively small cash awards.

To participate in ANCSA, individual Alaska Natives of at least one-fourth Indian, Eskimo, or Aleut descent were awarded one hundred shares of stock in both their regional and village corporations. If Native individuals could not identify a home village, they were given at-large shareholder status in the regional corporation; instead of village corporation stock, at-large shareholders received a per capita cash share paid by the regional corporation to its villages from the settlement fund. Settlement benefits in the form of stock were made available only to the approximately 79,000 Natives born before December 18, 1971. (See table 4.1 for regional corporation distributions and enrollments.)

Of the $962.5 million, the state of Alaska paid $500 million from the 12.5 percent royalty it receives from Prudhoe Bay oil. The federal government appropriated the remaining $462.5 million over a ten-year period. Money paid into the Alaska Native Fund by both parties was allocated to the regional corporations on the basis of their shareholder enrollments. During the first five years of the settlement (1971–76), 10 percent of the cash was distributed to individuals and 45 percent went to the village corporations. Thereafter all payments were kept by the regional corporations, and individuals have been paid dividends from corporate profits. Owing to poor corporate performance, the size and frequency of dividends has never approximated original expectations and in most cases has had a negligible impact on the annual incomes of shareholders.

Lands selected by Natives were divided between village and regional corporations. Although intraregional distributions were determined by village enrollment, the overall land allocation between regional corporations followed a much more complex formula.[4] Because village corporation lands included only the surface rights, the regional corporations control the subsurface rights to all Native-owned land within their boundaries. Imposition of the superintermediary power over the villages by the regional corporations has generated substantial conflict that has in a few instances been resolved by mergers.[5]

Comparisons between ANCSA and the Dawes Act suggest that in both

Table 4.1 Profiles of the Thirteen Native Regional Corporations

Corporation	ANCSA Entitlement (Land)	ANCSA Entitlement (Money)	Number of Share-holders	Number of Villages
Ahtna, Inc.	1.7 million acres	$13.3 million	1,074	8
Aleut Corporation	1.25 million acres subsurface rights; 52,000 acres surface rights	$40.1 million	3,124	13
Arctic Slope Regional Corporation	4.6 million acres	$46.5 million	3,710	8
Bering Straits Native Corporation	2.9 million acres	$79.5 million	6,271	17
Bristol Bay Native Corporation	2.2 million acres; 2.9 million sub-surface	$67.0 million	5,315	30
Calista Corporation	5.9 million acres	$165.0 million	13,193	56
Chugach Natives, Inc.	1 million acres	$24.0 million	1,881	5
Cook Inlet Region, Inc.	1.2 million acres surface and sub-surface rights; 1.3 million subsurface	$77.2 million	6,052	7
Doyon Ltd.	9 million acres surface and sub-surface rights, 3.5 million acres sub-surface	$112.3 million	8,905	34
Koniag, Inc.	1 million acres	$41.3 million	3,267	16
Nana Regional Corporation	2.3 million acres	$59.9 million	4,761	11
Sealaska Corporation	280,000 acres	$196.9 million	15,388	12
Thirteenth Regional Corporation	None	$45.6 million	4,435	None

cases economic motivations were the principal forces behind their enactment. In his study of allotment policy Otis argues that the most powerful motivating force behind the Dawes Act was "the pressure from land hungry white settlers" (Otis 1973, 3–12; Debo 1966). This assertion seems well corroborated by the historical evidence. In Alaska a land settlement was needed to secure a right-of-way for the construction of the Trans-Alaska oil pipeline. Hanrahan and Gruenstein have pointed out that the lobbying effort of major multinational oil companies was the major impetus for a congressional settlement. According to these writers: "The oil companies saw that the Native land claims stood in the way of a pipeline permit, but they soon realized that many of the Native leaders also favored development of oil and minerals on land they might receive under a claims act. Accepting the reality that the pipeline would not be built until the claims issue was settled, the oil companies joined with the Natives to lobby for a just lands settlement" (Hanrahan and Gruenstein 1977, 95). Although the usual procedure for resolving aboriginal land claims was to litigate these disputes before the United States Indian Claims Commission, the decision to seek a settlement from Congress was part of an effective confrontational strategy organized by a politically astute group of young Native leaders.

Like the Dawes Act, the Alaska Native Claims Settlement Act imposed time restrictions on the sale or dispossession of the award. In ANCSA Congress mandated a twenty-year protection period so that shares of stock in Native corporations could not be sold until December 18, 1991. Congress further stipulated additional provisions that prevent taxation of undeveloped Alaska Native lands.[6]

The important differences between these two pieces of legislation indicate a growing sophistication on the part of the government in dealing with Native peoples. In ANCSA the primary organizational form selected was the corporation. Despite many good reasons to the contrary, this basic institution of American business was chosen to become the assimilation vehicle for integrating Alaska Natives into the dominant capitalist political economy. For the most part, the basic goals of ANCSA and the Dawes Act appear similar, but the legal devices used to achieve these ends differ in that the Alaska Natives were artificially and temporarily brought together through regional and village corporations, whereas the Indians were handled individually.

PERMANENCE AND EXTENT OF OWNERSHIP

Although it is still too early to draw conclusions regarding the Alaska Natives' struggle to maintain control of the ANCSA land base, it is possible to make some reasonable extrapolations given the historical analogues to ANCSA and the likelihood of achieving uniform levels of corporate performance. Because the sentiments of the majority of shareholders are beginning to reflect major dissatisfaction with the corporations, there is very likely to be a major sellout of ANCSA corporate stock in 1991 even though a new law was recently passed to prevent such transfers.[7]

Even with the extensive legislative changes made to ANCSA, the probability of Native corporate failure looms on the horizon. Several of the regional corporations have encountered serious financial problems and are seeking legal relief through bankruptcy. The Thirteenth Regional Corporation is close to being dissolved. Another regional corporation is in receivership to a major California bank. Still another is being sued by its shareholders for various claims—including fiduciary irresponsibility.[8] In total, of the thirteen Alaska Native Regional Corporations created by ANCSA, fewer than half have reasonably good long-term probabilities of survival as Native-owned entities. Moreover, of the more than two hundred village corporations, fewer than one-third continue to prepare the necessary status and performance reports. Thomas Berger, a former Canadian judge who completed a two-year study of ANCSA for the Inuit Circumpolar Conference, found that many Alaska Natives fear that "through corporate failure, corporate take-overs, and taxation, they could lose their land" (Berger 1985, 6).

Currently ANCSA allows mergers between corporations, although it stipulates that there cannot be fewer than seven regional corporations. Because the law has been changed to give the regional corporations first right of refusal of stock put on the market, there is the growing possibility that regional corporations could execute a buyout program, thus leaving a smaller set of extremely powerful, closely held corporations. Given that the largest number of shareholders in any single corporation is under sixteen thousand and the smallest slightly more than one thousand, it is not difficult to envision such a scenario.

In essence these takeovers would transfer extremely valuable natural resources, along with controlling interest in Native Corporations, thereby allowing outsiders to pursue development projects that would not otherwise be allowed by environmental lobbies. The combination of natural resources plus political influence under a Native corporate banner provides powerful incentives for both inside and outside takeovers.[9]

A realistic view of ANCSA considers that the intended role of the regional corporations is to establish a developmental infrastructure and expend their initial cash awards to prepare the lands for other entities that will later reap the profits. When one considers that the probability of success for a government-financed, minority-owned business is very low, this seems likely. Realizations such as these have also prompted concerned corporate and community Native leaders to seek substantial changes through village sovereignty movements.[10]

EXTERNAL IMPOSITION AND INTERNAL RATIFICATION

Like the Dawes Act, the 1971 Alaska Native Claims Settlement Act also has a relatively long history that is usually traced back to the Russian/American Treaty of Cession (1867). But it was not until 1968 when oil was discovered on the North Slope that national attention was cast on resolving the Alaska Native land claims dispute (Arnold *et al.* 1978). Afterward the Alaska Native claims movement began to gather new types of support, leading to the formation of a statewide organization, the Alaska Federation of Natives (AFN).

Although it is impossible to establish a single motivation that completely explains the formation of the Alaska Federation of Natives in 1966, it is clear that concern for the protection of lands used for traditional subsistence was widely felt by most Native leaders. Another motivation, equally strong but articulated much less, was the expectation that the Alaska Natives were finally going to get an opportunity to cash in on the unlawful taking of their land. Between these two often conflicting and sometimes complementary forces, most Alaska Natives have expressed a strong determination to see that their lands are protected. One consequence of ANCSA corporations has been that some leaders have gravitated toward other personal concerns, with material gain being at the forefront.

Almost from its very beginning the AFN leadership seemed to insist upon corporations as a settlement vehicle. The reasons given for selecting corporations have to do mainly with the Natives' desire to reduce the bureaucratic control of the Bureau of Indian Affairs (BIA) (Federal Field Committee 1968). According to an attorney who represented the Arctic Slope Native Association during land claims negotiations, "The corporate idea was accepted because traditional tribal governments were seen as tools of the hated Bureau of Indian Affairs" (*Juneau Empire* February 28, 1984, 5).

The ANCSA final settlement was the result of numerous compromises that modified the bill so it lost much of its original intent (Berry 1975, 165–214). For example, one of the earliest bills submitted to Congress calls for dividing Alaska into seven Native regions according to a common heritage and forming a corporate structure "devoted to promoting the health, welfare, education, and economic and social well-being of its stockholders and their descendants." Compared with the final version of ANCSA, the Native-submitted bill emphasized a completely different set of responsibilities, going well beyond the business functions of corporations. Congress handled the differences in intent by calling for the creation of two types of regional corporations, one strictly for profit making and the second a nonprofit social service agency. Although these two entities were established with the belief that they would be equals, the fact is that for-profit corporations quickly became dominant.

Until fairly recent times most Alaska Natives lived almost entirely off the land by harvesting various species such as salmon, deer, moose, caribou, seal, walrus, and whales. These and other game animals provided the material basis for Native societies. The hunting-gathering way of life and the extremely harsh climate forced the evolution of culturally bounded small bands oriented toward cooperation for mutual survival. By contemporary standards life expectancy was low, and it was made even more uncertain by periodic famines. After the arrival of Westerners, less than a century ago for the most part, Alaska Natives gradually began to adapt new technologies to their hunting and fishing. Change came more quickly as a result of BIA boarding schools and the military buildup during the Second World War. By the late 1960s the rural subsistence economy of Native Alaskans had been seriously affected by the technological diffusion and by the expansion of social welfare programs stemming from the War on Poverty and President Johnson's Great Society. Still other causes for more pronounced social and economic change vary directly with the acceleration of geographical mobility and the escalation of state government spending derived from oil revenues. Similarly, Native political power arising from the lands settlement has strongly influenced a broad range of activities that have profoundly altered village life.

Given the limited understanding of ANCSA and major differences between Western and traditional Native views of landownership, it is possible to argue that ANCSA was not a legitimate expression of Alaska Natives' self-determination. The numerous discontinuities arising from ANCSA, especially with regard to the effects of external forces on the organic texture of Alaska Native communities, indicate that the probable intent of ANCSA was not to preserve Alaska Native landownership. As Ferguson, a past president of AFN, argues: "I believe the Natives who were shooting for this won a fair settlement, but the people who devised the bill knew that this was very complicated and that failure was likely" (Lenz 1981, 19).

Unlike the Dawes Act, where external imposition is more pronounced, the manipulation in ANCSA is less overt. And though there was an internal ratification by the six hundred members of the AFN, in truth most Natives in Alaska had little appreciation of the complex terms of the land claims that would be passed.

SOCIAL AND ECONOMIC EFFECTS

As with its historical predecessor, the Dawes Act, a major effect of ANCSA has been to interject a high degree of divisiveness among the recipients. When one looks at the organizational dimensions of the ANCSA, it is apparent that a concerted effort was needed to bring regionally and culturally different Native groups together to form the statewide organization called the Alaska Federation of Natives. During the intense lobbying that preceded ANCSA passage, the AFN was the unifying entity that helped resolve differences between groups (Lantis 1973, 99–118).

Almost immediately after the passage of ANCSA, the unity of Alaska Natives fragmented. Certain sections of the act created divisions in the control and use of the land and cash settlements. Although the level of structural differentiation as reflected by life-styles, leadership orientations, and economic opportunities differs among regions, the breakup in the cohesiveness of the Alaska Native community and the attendant effects of social disintegration has been exacerbated by value conflicts between traditional usufruct property ways and the new focus on private property. As Garber (1984) poignantly observes:

> With little faith in federally supervised property, Native leaders opted for parity with non-Natives in land tenure and saw in ANCSA an alternative through stock ownership in landed corporations. With thousands of years of experience in holding communal property and relatively little experience living with private property, it's no wonder that conflicts arise between the new land tenure under ANCSA and the older notions of communal property. . . .
>
> Indians in the south faced similar conflicts when their reservations were first broken up by allotment and reconstructed under the Indian Reorganization Act.

Three levels of conflict that deserve closer examination deal with the regional corporations' relationship to each other; the relationship between villages and their regional corporation; and the relationship of the shareholders to their corporations.

Section 7-i of ANCSA calls for a 30 to 70 percent distribution of net revenues between a regional corporation generating income on timber sales or subsurface mineral developments and all of the regional corporations. According to the negotiated terms of a costly legal dispute between ANCSA corporations, a regional corporation is to receive 30 percent of its net revenues from timber and subsurface development and a per capita share of the remaining 70 percent, which must be divided with all other regional corporations. This requirement has been extremely divisive, causing financially profitable corporations to dilute their revenues and making additional investments in new projects uneconomic given the risks associated with full liability but limited returns.

Because ANCSA gave the regional corporations certain superintermediary powers over the village corporations within the region, there have been numerous legal conflicts dealing with developments on lands used for traditional subsistence purposes. Admittedly, there are numerous instances where, for various reasons, Alaska Native Regional Corporations have made extraordinary efforts to accommodate the concerns of village corporations within their boundaries. Nevertheless, there are cases where the protective attitude has been eroded by the profit motive or not extended to Native villages outside the region. Examples where business activities may have a serious impact on the environmental quality, health, and subsistence resources of villagers include oil drilling in the whaling grounds of the Beaufort Sea; lead and asbestos mining projects that have been proposed in the NANA and Doyon regions; and clear-cutting of prime wildlife habitat in southeastern Alaska. Each of these examples is likely to impair rural villages in a number of ways. Still, the most pronounced effects of regional corporations are economic in nature. Although many rural Alaskans have been afforded basic services in increasing numbers, most of the villages have not developed the necessary economic base to support these services in the face of dwindling state revenues. Unfortunately the reduction in state grants and other forms of support has not been met with intensified efforts to identify local economic activities that promote village employment.

At the same time that the regional corporation had final authority to approve the development plans of its village corporations, regional corporations were given control over subsurface resources. In addition to the competition for the land resources between regional corporations and villages, the competition for qualified Native leadership also exacerbates the tendency for regional corporations to gain overpowering control of the villages, limiting the range of development opportunities undertaken regionally (Anders 1983, 555–75).

Although some regions have to a large extent reduced these internal conflicts through mergers with village entities, in cases such as the Koniag Regional Corporation the rivalry between villages and the regional corporation led to lawsuits that have practically bankrupted the corporations involved. Such patterns substantiate the frequent observation that the principal ANCSA beneficiaries have been the state's attorneys.

Individualization of communal peoples' land is a powerful process, whether under the General Allotment Act or the ANCSA. Stavehagen (1975, 112) argues that it produces "a polarization characterized by the existence of many people who hold very little land and a few people who control a great deal of land. This concentration of landownership has determined the relationship among social classes in the countryside and has strongly influenced social stratification systems." What ANCSA did in effect was to create various classes of Alaska Natives as the Dawes Act and the Indian Reorganization Act had done before. At the top of present Alaskan Native society are the handful of Alaska Natives who have moved into privileged positions in an institution, such as the regional corporation or nonprofit social service agencies created to serve underprivileged Alaska Natives. These generally urban Alaska Natives are employed in relatively high-paid jobs and receive the status and prestige that elite positions offer.

The second and most numerous class of Alaska Natives comprises members of a rural population that faces much more limited employment prospects. They by and large have received miniscule benefits from ANCSA and have been only marginally affected by the corporations' activities. Their primary occupation is still in areas of subsistence harvesting of natural resources, with supplementary cash coming from seasonal jobs or public sources. As Arnold (1982, 13), a longtime ANCSA observer, puts it: "Despite the cash payments received by Eskimo, Indian and Aleut shareholders or the value of shares they won in their corporations, and the prosperity and political influence that some possess, the lives of most Natives appear to be little changed by the most extraordinary aboriginal land claims settlement in American history."

A third class of Natives are those who face the real prospect of being completely excluded from any ANCSA benefit. Alaska Natives born after the final enrollment currently receive stock only through probate and are not eligible to vote in corporate elections until 1991. The recently passed amendment allows corporations to issue special classes of stock in an attempt to include younger Natives as shareholders but there has been little discussion of the effects of issuing new shares on the existing shareholders assets. Still, the tensions created between Natives born before 1971 and those born after the land claims act was passed will not be easily resolved.

To restate the theme of this chapter, ANCSA, like the Dawes Act, has unleashed a storm of competing interests and conflicting expectations on Alaska Natives who (in varying degrees) had maintained communal bonds through community decision making, cooperation, and sharing. By breaking up the land base, ANCSA factionalized communities, terminated tribal relationships, and interjected a dynamic of individualization and conflict that divides people of common culture. It moved the pattern of land use one step further toward the "private property" concept and began the privatization of common resources. Traditional native culture is survival oriented and kinship based, but suicide and accidental death rates indicate there has been great destruction from the shredding of the community fabric that has held Alaska Natives together. It is apparent that this is a situation where a more powerful entity (the federal government) is responsible for imposing a pattern of self-selection and limited success by those with the least investment in traditional culture. The likely result of ANCSA will be a consolidation of ownership of valuable land and natural resources in the hands of individuals who are the most likely to see those resources developed, despite the environmental or sociocultural effects.

CONCLUSION

The thesis of this chapter is that the Alaska Native Claims Settlement Act conforms closely to the policy of allotment in that it has effected the breakup of tribal organizations, engendered the formation of self-selected elites who have exploited the new opportunities for profit making, and set the pattern for new land uses that will destroy the natural resource base of the old subsistence economies.

In the Alaska Native situation we see the competition between policies offering vastly different possibilities. Tribal governments created by villagers who have grown increasingly dissatisfied with ANCSA seek to convey their lands to sovereign tribal entities recognized by the federal government in the same way that reservations are recognized as incorporated entities. The growing strength of the statewide challenge to the dominant establishment-oriented position of AFN has resulted in a major schism. Groups such as the Alaska Native Coalition, the principal organization of village Natives, are calling for changes that will ensure the permanence of the land award through transfers to village tribal entities. Certainly the sovereignty movement poses a serious challenge to the original intentions of ANCSA. But given the dominant assimilationist role of the ANCSA regional corporations in the existing power structure, it is unlikely that a serious effort will be made to radically alter the present organizational framework.

It is very difficult to assess the long-term effect that ANCSA will have on Alaska Natives. The land claims onslaught is only a part of the disruption, which has included alcohol, drugs, the loss of traditional social controls, and the introduction of the problems of the whole of Western culture transmitted in Alaskan cities and rural villages. Certainly a great many things are lost beyond reconstruction. New generations of Alaska Natives brought up in the wake of welfare dependency have come to associate economic development with increased public funding. The entire concept of a sustainable and continuing community—with a flexible organization that could respond to conditions imposed by the availability of game and community needs—seems to have been among the first victims. Since the establishment of permanent villages, some Alaska Natives (particularly the Western Eskimos) have proved extremely vulnerable to the most corrosive influences of whites.

One lesson that can be learned from the American Indian experience is that Native people are more resilient than modern institutions. Despite allotment, termination, and relocation programs, plus a number of other federal policies designed to deal with the Natives by integrating them into the larger society, many have resisted. Alaska Natives are attempting to achieve something that has not yet been possible in other cases. In their efforts to avoid the inevitable usurpation of their traditional lands, they successfully lobbied for legislation providing for a land base and corporations. The twenty-year protection period on the sale of stock was the mechanism that was to allow Native leaders to prepare their people so that by 1991 the shareholders could have a more or less equal voice in their self-determination. This time allowance has proved very short indeed.

Not disregarding the continuing political acumen of Alaska Natives in effecting amendments and favorable changes in the laws relating to ANCSA, I contend that a fundamental incompatibility still exists between the functional aspect of the corporations (at least in the western conception) as a vehicle for assimilation and the great majority of Alaska Native villagers. It seems clear that the corporations were instituted in order to acculturate a Native elite through which Western influences and values could be communicated to others. Of course, substantial differences in history and cultural attitudes must be taken into consideration, but it appears, based on the American Indian experience, that only a few of the ANCSA corporations are capable of resolving deep internal conflicts arising from the limited applicability of this model.

NOTES

1. Congressional review of ANCSA is required under section 23. This review process was delayed until early 1988.

2. Under section 14(h) Congress set aside 2 million acres for individual allotments of 160 acres or less that would be registered within four years after passage, but it canceled opportunities for Alaska Natives to receive individual allotments under the General Allotment Program. Lands remaining after the application deadline were conveyed to regional corporations based on their shareholder enrollment. Section 14 (C1) required village corporations to convey to individual Natives title to surface estate for tracts used as primary places of residence, business, or subsistence.

3. Villages that voted to retain their former reserve lands and status include Venetie, Elim, Gamble, Savoonga, Tetlin, Klukwan, and Arctic Village. These villages received land taken from the public domain.

4. Regional corporations were entitled to receive 16 million acres under a complicated formula that considered not only the number of shareholders within the region, but also historical use and occupancy. Participating village corporations had three years to select from three to seven townships (69,120 to 161,280 acres). Land already in private ownership could not be selected, and additional set-asides were made for national parks and defense installations.

5. Mergers between villages and regional corporations are allowed; however, ANCSA section 7B requires that there be no fewer than seven regional corporations.

6. These and other changes were effected through the 1980 Alaska National Interest Lands Conservation Act, PL 96-487.

7. The Alaska Federation of Natives proposed a series of amendments to ANCSA that were recently passed by Congress. These amendments include extending land bank provisions to prevent taxation and allowing the Native shareholders of each corporation to ban the transfer of stock to non-Natives. They also grant voting rights to Natives receiving stock through inheritance, allow the issuance of special classes of stock to elders and Natives born after 1971, and allow the corporations to purchase stock from shareholders before 1991.

8. This controversy deals with Koniag Corporation's attempt to gain control of valuable timber lands on Afognak Island (*Alaska Business*, September 1983, 27).

9. A good example of the influence that the Alaska Native Regional Corporations can bring to bear on both the state and federal governments is the Red Dog lead/zinc mining project. In a joint venture with Cominco International, NANA, a Kotzebue-based corporation, has been able not only to arrange land exchanges that promote the construction of a rail line to transport the ore to a port, but also to secure over $200 million in long-term, low-interest state loans. Furthermore, NANA succeeded

in getting 2.1 million acres transferred from the North Slope Borough to provide the tax base for the formation of their own borough government.

10. The AFN lobbying effort focused on the 1991 considerations regarding the sale of stock by Native shareholders. This approach was strongly challenged by other Native leaders seeking to reconvey lands to federally recognized tribal groups. An important related facet of the rival sovereignty movement is the transfer of game management authority from the state to these IRA governments.

REFERENCES

Anders, Gary. 1983. The role of Alaska Native corporations in the development of Alaska. *Development and Change* 14:555–75.
———. 1985. A critical analysis of the Alaska Native land claims and Native corporate development. *Journal of Ethnic Studies* 13:1–12.
Arnold, Robert D. 1982. '71 settlement in retrospect. *Alaska Native News*, November.
Arnold, Robert D., *et al*. 1978. *Alaska Native claims*. Anchorage: Alaska Native Foundation.
Berger, Thomas R. 1985. *Village journey: The report of the Alaska Native Review Commission*. New York: Hill and Wang.
Berry, Mary C. 1975. *The Alaska pipeline: The politics of oil and Native land claims*. Bloomington: Indiana University Press.
Debo, Angie. 1966. *And still the waters run*. New York: Gordian Press.
Federal Field Committee for Development Planning in Alaska. 1968. *Alaska Natives and the land*. Washington, D.C.: Government Printing Office.
Fuller, Lauren L. 1976. Alaska Native Claims Settlement Act: An analysis of the protective clauses of the act through a comparison of the Dawes Act of 1887. *American Indian Law Review* 4:269–78.
Garber, Bart. 1984. 1991 Roundtable discussion paper, Alaska Native Review Commission. Photocopy.
Hanrahan, John, and Paul Gruenstein. 1977. *Lost frontier: The Marketing of Alaska*. New York: W. W. Norton.
Langdon, Steve J. 1982. Alaskan Native land claims and limited entry: The Dawes Act revisited. Paper presented at the American Anthropology Association annual meetings, December 4–7.
Lantis, Margaret. 1973. The current nativistic movement in Alaska. In *Circumpolar problems: Habitat, economy, and social relations in the Arctic*, London: Wheaton and Exeter. Vol. 21, 99–118.

Lenz, Mary. 1981. Native Claims Settlement Act—Was it meant to fail? *Tundra Drums* 9:19.

McBeath, Gerald A., and Thomas A. Morehouse. 1980. *The dynamics of Alaska Native self-government*. Lanham, Md.: University Press of America.

Otis, D. S. 1973. *The Dawes Act and the allotment of Indian lands*. Norman: University of Oklahoma Press.

Price, Monroe E. 1975, 1976. Regional-village relations under the Alaska Native Claims Settlement Act. *UCLA-Alaska Law Review* 5:1, 2.

Rogers, George W. 1969. Party politics or protest politics: Current political trends in Alaska. *Polar Record* 14(91): 455–58.

Stavenhagen, Rodolfo. 1975. *Social classes in agrarian societies*. Garden City, N.Y.: Anchor Books.

Chapter Five

The Impact of Colonialism
on Land Use in
Central and Southern Africa

Robson Silitshena

The colonial experience of peoples of central and southern Africa in the areas of land use, land management, and settlement in many ways resembles what occurred in the Great Plains and Alaska. That experience is encapsulated in the remark of the Tswana chief who complained that his people were not afraid to tell him that the white people, not he, were now master (cf. note 3). But more than power was at stake as the loss of land and restriction of movement consequent upon the onset of colonial rule created declining living standards for Africans and set off a chain reaction that included labor migration to the towns and the degradation of the natural environment to an appalling condition.

The peoples discussed in this chapter live in South Africa, Swaziland, Lesotho, Botswana, and Zimbabwe,[1] and most of the land was once grazing land where these mixed agriculturalists/pastoralists kept their large herds. The altitude ranges from sea level along the coastal plains to high mountain areas in Lesotho and Zimbabwe (over 3,000 m; map 5.1). Much of the interior is part of the African plateau, whose highest parts in Swaziland, South Africa, and Zimbabwe, referred to as the highveld, have a pleasant climate (1,500–2,500 m). Except for the Western Cape, much of the area receives some rainfall in summer, though it is also a time of very high temperatures. The cool, dry season runs from May to August, and in some places the temperatures can drop below 0°C and frosts are frequent. The rainfall generally

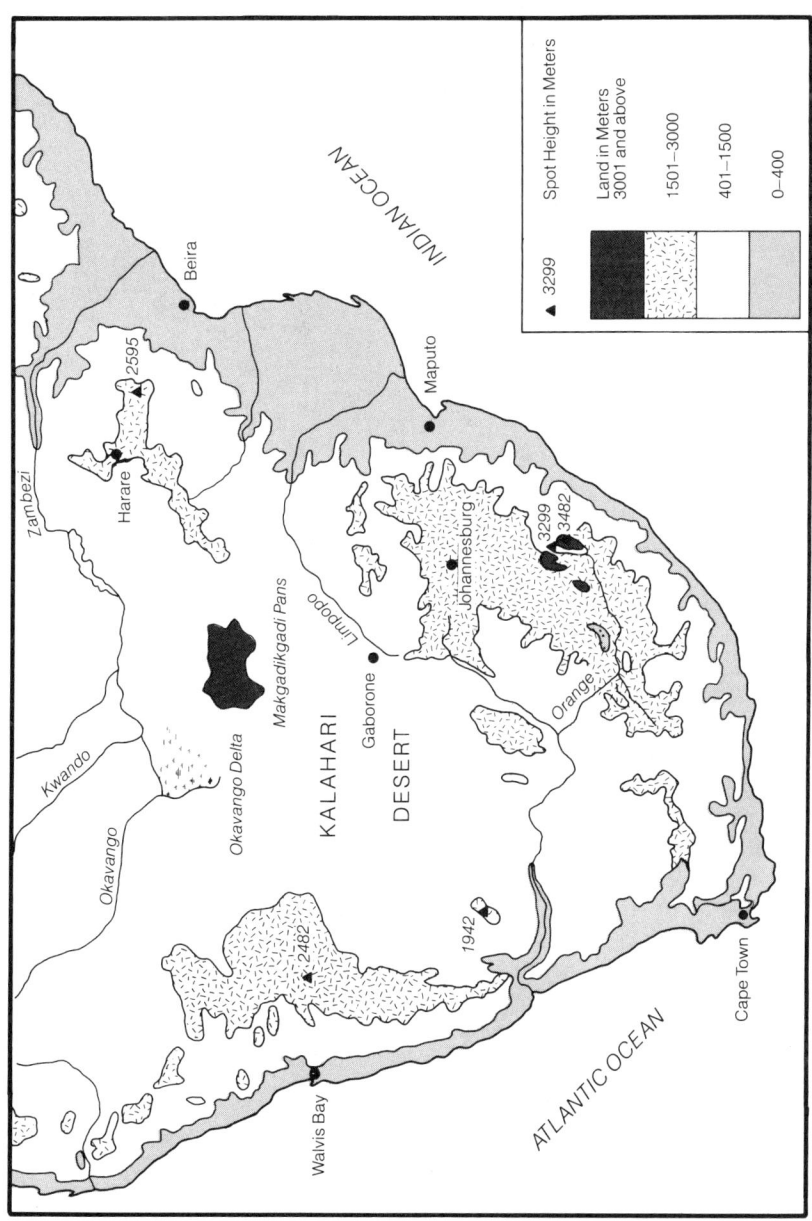

Map 5.1 Southern Africa: physical features.

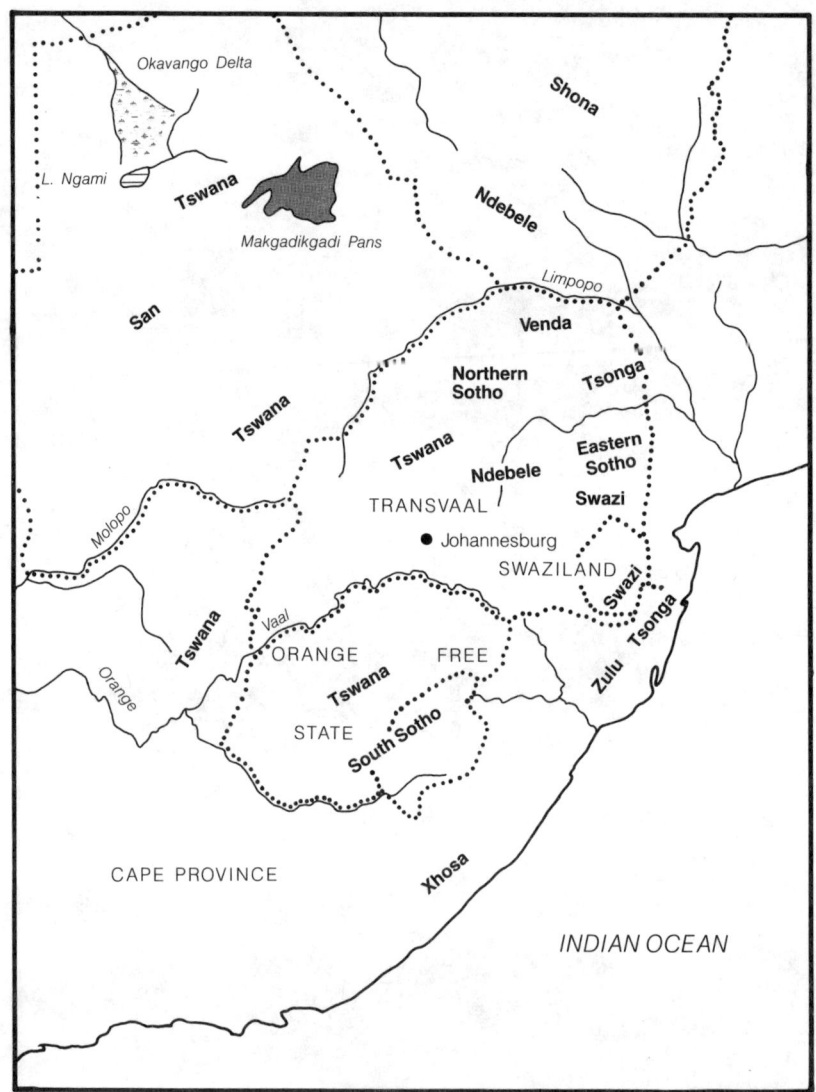

Map 5.2 Distribution of some ethnic groups in southern Africa.

decreases from the east coast (about 1,000 millimeters) to the west, where in some places in the Kalahari desert rainfall drops to 250 millimeters. However, rainfall effectiveness is reduced by the high rate of evapotranspiration—double the amount of rainfall received—and hence much of the area, especially the western part, is drought prone. Research has established that rainfall tends to occur in roughly twenty-year oscillations during which years of good and poor rainfall alternate (Tyson 1979)—similar to the rainfall cycle of the Great Plains.

The soils and vegetation in the area under consideration roughly correspond to the climatic belts. The western, drier parts are characterized by shallow, less developed Kalahari sandy soils having little humus content, while the soils of the eastern areas are generally better developed and more productive. Various kinds of savanna predominate, but other main vegetation types are also present—the grasslands (veld) of South Africa and the small Mediterranean area of the Cape. In short, much of the area is a somewhat warmer version of the Great Plains, a generally hostile environment susceptible to drought. But in contrast to much of the Plains, the soils are not excellent and vegetation is sparse. Many areas are currently threatened with desertification on account of intensifying overgrazing.

The region is inhabited by people belonging to various ethnic groups (map 5:2) and falling into six main linguistic groups—Sotho, Nguni, Tsonga, Venda, Shona, and San—which can be further divided into subethnic groups. The Nguni, comprising the Zulu, Xhosa, Swazi, and Ndebele, were originally concentrated on the eastern margins of present-day South Africa. The Ndebele, most of whom are now in Zimbabwe, left this area during the first half of the nineteenth century as a result of upheavals that followed the rise of the Shakan empire (Omer-Cooper 1966). The Nguni were traditionally organized according to clan and lineage (Wilson 1969a, 116–30), the latter consisting of all the descendants of a common ancestor in the male line. Lineages varied in span from three to six generations, and the members of a lineage often lived together in one homestead, whereas a clan was composed of a number of lineages that claimed descent from a common ancestor and did not intermarry. Beyond the clan existed the local groups, confined to localities, which in turn made up chiefdoms. In any local area members of a clan predominated, but an area always included some people from outside the clan. A chiefdom was a political unit occupying a defined area under an independent chief, though some of the chiefdoms were large and were subdivided under subordinate leaders. Chiefdoms were periodically reduced in size by frequent splits between the followers of the first wife's son and those of the great wife's son (Wilson 1969a, 116–30), one way of matching the population to the resources.

The Sotho group—for example, the Tswana, the Basotho, and the Pedi—on the other hand, lived in the interior elevated, more arid parts of southern Africa. In contrast to the Nguni, they not only preferred kin marriages, but also lived in nucleated settlements (Kuper 1975, 67–81) and organized themselves into chiefdoms with a ruling aristocracy that provided the chief and other powerful officeholders. The chiefdoms were divided into administrative subdivisions, the most important being the ward—the administrative unit within which rights in fields and pastures were allocated. Among the Sotho, power was monopolized by the aristocracy, the patrilineal descendants of the first rulers among whom the competition for power took place, most particularly during periods of succession. The defeated group would eventually hive off and settle somewhere else, thus relieving pressure on local resources. Below the wards were family groups built as Kuper describes it, "about a core of closely related homestead heads, typically brothers, sons, or brothers' sons of the group head," lineage segments that typically excluded some agnates in the same categories and included "matrilateral and affinal kin who are not agnates of the core group" (Kuper 1975, 71). The base was a family, occupying a homestead. A typical homestead comprised a nuclear or polygynous family at some stage of the developmental cycle. It might include other kin, usually a divorced or widowed woman. Sotho society, particularly the Tswana branch of it, was heavily stratified into four levels, the top including the aristocracy and the bottom the San and other minor tribes (Schapera 1953, 36–37). The other people of the region were equally diverse in cultural practice and organization. The mixture of activities they pursued might be compared to those of the groups on the eastern Great Plains such as the Omahas or Pawnees, except that they had herds as well as practicing hunting and agriculture.

As among other African societies, land was generally communally owned by the Nguni, the Sotho, and the other groups in the area (Schapera 1943, 1955). Individuals had usufruct rights to it, though it was often held by the chief as a trust from the entire community, and he in turn allocated it to individuals or groups for various uses, a role he usually delegated to the headman below him. The land for residential and arable uses was often allocated to a household and became de facto private property that could be inherited. However, if the land was not used for a time, it could be repossessed and allocated to someone else. Grazing land was commonly communal property, with all members of the community grazing their livestock in the same area. Among the Tswana both grazing land and arable land were originally allocated by ward (Schapera 1943). Land had no market value and could not be sold. The precolonial states were small scale, and boundaries were usually fluid. Because populations were small relative to available land, movement was the main method of reducing pressure on resources.

The backbone of the Shona economy was typically livestock farming, supplemented by subsistence cropping, trading, hunting, and gathering. Population movements to relieve overpopulation were common (Beach 1977, 41). Hoe crop agriculture was precarious because of pests, but drought was the major problem faced, since there was about one chance in five that a year would bring subnormal rainfall (Beach 1977, 43). Indeed Beach reported that "traditions of movements show that at times the agricultural system simply collapsed and people had to fall back on gathering and hunting, or flee to other areas not so badly hit" (1977, 43). The missionaries and early nineteenth-century travelers observed a similar situation among the Tswana (Hitchcock 1979, 91–98).

Livestock, particularly cattle, thus played an important role in the economies of these societies as the source of milk, meat, and prestige, as a medium of exchange, and as the means to pay bridewealth. However, cattle were killed only on special occasions: "In the last resort," Beach writes, "the cow was insurance against the failure of the crops, and this lay at the root of its economic importance, but it acquired so much social importance that the main aim of every Shona farmer was to build up his herd" (1977, 46).

The mixed grazing/agricultural societies of southern Africa were never self-sufficient in everything and have for hundreds of years engaged in local, regional, and even international trade. When crops failed it was necessary to purchase grain from other societies that had surpluses. Other items such as iron and salt, whose distribution was restricted to certain areas, were traded among the local people, and items of external trade, mainly with the Indian subcontinent via the east coast, included gold, ivory, and copper (Beach 1977, 46).

Studies of land use and the management of space reveal a contrast between the Sotho system and that of the other societies. Sansom (1974) has contrasted the dispersed settlements among the Nguni with the nucleated settlements among the Sotho. The dispersed Nguni settlements were characterized by the concentration of all land uses in one place on account of *"small-scale repetitive configurations that contained a variety of natural resources"* (Sansom 1974, 140). In contrast, the relatively uniform plant ecology the Sotho confronted militated against concentrating all land uses in one area. Sansom argues that the terrain here was such that

> a variety of resources is less frequently contained in small . . . convolutions of Zulu hill country. On the inland plateaux one is often confronted with large expanses of relatively uniform country. To move from one type of plant to another, or to find different soil types, one must travel over larger distances. There is a general problem of finding a constant water supply and water resources are often far apart. Because people need to exploit variations of terrain, they must range over an extensive area. To accommodate a ranging and open strategy, the tribal territory replaces the district in supplying the self-contained area in which the variety of its inhabitants' requirements will be satisfied. (Sansom 1974, 142–43)

And so the Sotho traditionally lived in large nucleated settlements, some of them estimated by nineteenth-century European travelers to include 20,000 people. Since every man had to have a home in the town (Schapera 1943, 59), the fields extended a considerable distance from the edge of the towns (map 5:3), and cattle were kept at even more distant grazing stations, called cattleposts, selected for the quality of their grazing and accessibility to water.

Most of these societies offer little evidence of conscious efforts at managing natural resources. The main strategy for balancing resources with population was movement, but this may be the best one in tropical countries whose biomes are characterized by low productivity (Hitchcock 1982, 17). However, some Tswana groups practiced more deliberate resource management. Among the Ngwato and Ngwaketse, for example, some grazing areas were allocated to groups of wards and placed under an overseer, who controlled grazing and ensured that households were not grazing their cattle too close together (Hitchcock 1982, 4; Gulbrandsen 1980).[2] The Xhosa practiced rotational grazing, described as follows by Monica Wilson: "Areas were set aside by the senior chief, and also by district chiefs, for cattle posts

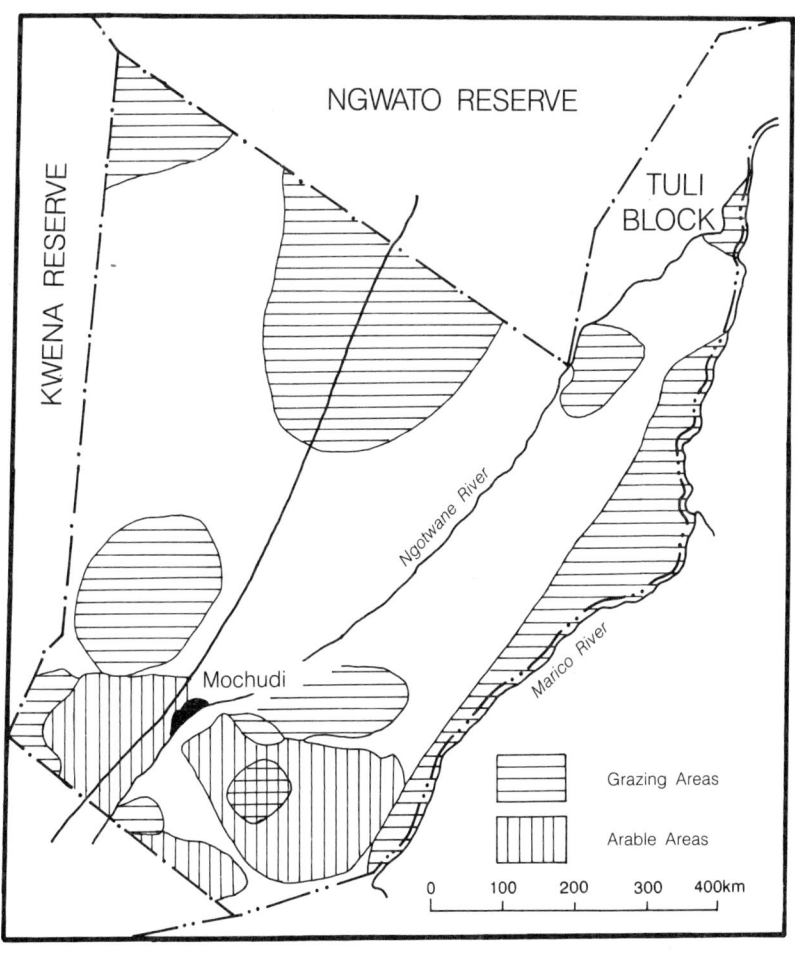

Map 5.3 Land use in Kgatleng (Botswana), 1935.

to which stock were sent to graze at certain seasons, and use was made in this way of different types of pasture—sweet and sour veld as it is called in South Africa. There was a regular practice of transhumance, sending stock to the mountains in the summer, or to sourveld patches near the sea, but that changed after the Xhosa lost much of their land to whites, and themselves increased in numbers" (Wilson 1963, 381).

The colonial period that altered these resource-management patterns, though its full impact took time to be felt, extended for more than eighty years in most of this region and still goes on in South Africa and Namibia. Its negative effects on land use and management have been major factors in the underdevelopment of the societies of the region. Agricultural production has declined mainly as a result of policies that discriminated against African farmers, and the only alternative in most instances has been labor migration to the new mining and industrial nodes in order to cope with increasing population and rising expectations. Although the division may be artificial, it is useful to distinguish between the short-term and the long-term effects of the colonial period.

In assessing short-term effects, one must understand that European contact with the peoples of southern Africa goes back several centuries to the early Portuguese voyages to India. It became more intensive for some communities after the Dutch landed at the Cape in the seventeenth century, and European penetration into most of the interior took place about the middle of the nineteenth century, mainly as a result of the scramble for Africa. The settlement of European-based peoples in the semiarid regions of southern Africa at the expense of the indigenous peoples of the area took place at about the same time as European-based peoples were seizing the Great Plains from Native Americans. As they advanced into the interior, they grabbed the land from the Africans. For instance, there were struggles between Africans and the Boers as the latter proclaimed their republics on African-owned land in the interior. By the end of the nineteenth century the map of present-day southern Africa had been drawn as the scramble reached its crescendo. As in other parts of Africa, the various ethnic groups were chopped up by the new boundaries, divided among various countries.

Although the early encounters between Africans and Europeans may have been marked by hostilities, both societies also benefited from these exchanges. Clusters of European population offered the African farmers a market opportunity. For instance, after the discovery of diamonds in Kimberley, Basotho "vigorously responded to the mining camps' demand for food supplies. They exported vast quantities of grain" (Lye and Murray 1980, 137). Even in Southern Rhodesia, up to about 1920, the food needs of the local European community, preoccupied with a search for minerals, were met by African farmers. Indeed, it was more profitable for Africans to stay on the land and farm than to work for wages (Mason 1958, 227; Arrighi 1967). In Southern Rhodesia in 1903, European agriculture accounted for less than 10 percent of total marketed output (Riddell 1980, 6). Thus the first stages of the European incursion were marked by a concentration on minerals, not on minerals *and* land as in the white conquest of the Great Plains.

The commercialization of agriculture and the subsequent loss of land and the chiefs' power affected social and spatial structures. The Tswana in South Africa started to scatter from the towns to live close to their fields (Pauw 1960, 49–76; Dachs 1975).³ The earlier nucleated settlements disappeared and now remain only in Botswana (Silitshena 1983). The power over the spatial organization of societies that the chiefs once possessed had emanated from their control of land. As Werbner has observed:

> The span of a tribal ruler's authority was relative. It increased the more he was
> able to provide and allocate the resources which his people needed to meet ex-
> treme hazards in production, in raising of their crops, and in the herding of
> their livestock. . . . A paramount chief had to control such productive re-
> sources as would allow him a major concentration of subjects in a great village,
> and make him secure against its dispersal. . . . There was a systematic relation
> between paramount rank, settlement in a concentrated, centrally sited village
> and a specialized mode of production in zones set aside for pasture and, sep-
> arately, for cultivation. (Werbner 1971, 34)

Though the short-term effects of European occupation may not have seemed drastic, the long-term ones were, especially because of the creation of reserves exactly like those created for American Indians on the Plains. With the European occupation of the region, the land grab that started earlier was intensified and later followed by the exclusion of Africans from the areas that were now declared European owned. In Southern Rhodesia, Johnson notes, "land grants were given out very freely—without much regard" to the Africans (1968, 37). European estates varied in size from 2,400 hectares (6,000 acres) to 12,000 hectares (30,000 acres) (Floyd 1962, 98). The move to create reserves was supposedly to protect the Africans from loss of land. A better explanation was given by N. H. Wilson in 1923 when he observed: "The objects of our native policy in respect to these two phases of the native problem will therefore be two-fold. Firstly, the full economic development of the native as a unit of the State; secondly, the development of the native in such a way that he will come as little as possible in conflict or competition with the white man socially, economically and politically" (Wilson 1923).

In Southern Rhodesia the creation of African reserves began in 1894. By 1914, 8.4 million hectares (21 million acres), mostly in less favored natural regions, had been set aside for this purpose, and the process continued almost up to the end of the colonial period. Similar developments were taking place in other parts of the region. African reserves had already been created in parts of South Africa (Shillington 1985, 226–28). In 1913 the South African government set aside 9.5 million hectares for African use, an amount at that time "already inadequate for African needs" (Christopher 1982, 57). The total area of African land was, after long delays, increased to 15.5 million hectares or 13.7 percent of the total area of South Africa for 75 percent of the population. On the effect of these South African land laws, Lye and Murray have observed:

> Prior to 1913, their reserved lands provided the Africans with a base for economic and political life, but did not confine them. The land laws of 1913 and 1936 closed their right to own lands elsewhere. Territorial partition became the basis for development of South Africa. Africans could enter the white man's areas only on sufferance and squatters could be expelled. The Natives' Land Act of 1913 explicitly limited land for Africans as a means of assuring an adequate labour supply to the European farmers and industries, and to eliminate competition with the poor-white class to acquire land. The preservation of the reserves assured the Africans of a place where they could perpetuate their own culture, but it was not adequate to support them. (Lye and Murray 1980, 76)

The parallels with the reservation system provided for America's Indians are obvious.

Similar situations obtained in areas that came under the indirect rule of the British. Some groups—namely, the Tswana in present-day Botswana, the Swazi in present-day Swaziland, and the Basotho in present-day Lesotho—asked for British protection during the scramble to preserve their land, much of which they had already lost. The Swazi had their country taken up by concessions and have had to buy back some of the land (Parsons 1982, 221–22). The Basotho were left with a country that "was too small and too deficient in natural resources to sustain its population" (Thompson 1969) and had to survive by selling their labor outside their country; and parts of the better-watered and relatively more fertile eastern strip of Botswana were set aside for European occupation. Hence land still is a major talking point in the politics of the North East district, where there is an acute land shortage; worse still, two-thirds of Botswana is the Kalahari semidesert. Indeed, even in the protectorates Africans lived on reserves.

In the course of the nineteenth century, the economies of these territories were incorporated into the mercantile capitalism based at the Cape, which subsequently made them appendages of the South African economy and forced the selling of cattle and crop surpluses in the South African markets. The collapse of the cattle market between 1902 and 1910–11 put a severe strain on some of the local Botswana economies (Palmer and Parsons 1977), as did the imposition of the notorious minimum-weight restrictions on imported livestock (Ettinger 1972, 21–29). The incorporation of the protectorate economies into that of South Africa seemed logical in view of the expectation that they would eventually become part of South Africa. The point here is that Africans in the protectorates did not fare any better than those in South Africa and Southern Rhodesia, since they were subject to the same general forces. As Lye and Murray have observed, "The failure of peasant self-sufficiency must be examined against the background of a larger system of production. This demanded, not that Africans grow sufficient food to preserve their economic independence, but that they contribute their labour to the 'white' economy—to white farmers and mining companies in particular" (Lye and Murray 1980, 138).

The British administration connived, through such mechanisms as taxation, to ensure a free flow of labor from the protectorates to the mines (Massey 1978, 95–98). Cheap labor was perhaps the most important commodity demanded by the South African economy (Legassick 1977).

The initial spurt of commercial agricultural development in African areas was short-lived, since once the Europeans became interested in farming the state threw its full weight behind them. In the early days the rural white community formed the bulk of the electorate, and the physical separation of the races following the introduction of the land acts made it possible for white governments to pump considerable capital into the white areas for the development of an economic infrastructure and the provision of credit, and this capital played a key role in the development of European agriculture (Arrighi 1970; Houghton 1976, 65–68), making it possible for even inefficient producers to remain in business (Riddell 1980, 5). Even the marketing system favored white farmers at the expense of black, since the prices paid to white and black farmers were not identical (Silitshena 1972, 33–37).[4]

European or commercial agriculture therefore played, and continues to play, an important role in the economy of the various countries of the area. It creates at least 80 percent of the agricultural production and over 90 percent of total marketed output, and it contributes handsomely to the gross domestic product and to foreign-exchange earnings. It is a major employer, accounting for at least 40 percent of the labor force in the formal sector, provides most of the national food requirements, and supplies a considerable proportion of the raw material used in manufacturing.

Traditional African agriculture contrasts with European-based commercial agriculture, which tends to be capital intensive and works on a large scale. In the African areas of South Africa the plots per family are small, about 2.7 hectares. Access to credit is negligible. Use of inputs such as pesticides and chemical fertilizers is rare in traditional subsistence production, though people do allow cattle to graze in the fields after harvest, fertilizing the soil with dung. Although crop rotation is not widespread, people do practice interplanting, adding nitrogen-fixing crops such as beans that help maintain soil fertility. Marketing infrastructure is poorly developed in the rural parts of southern Africa, and transport costs are high. It is not surprising, therefore, that, as Christopher (1982, 61) notes, "Migratory labour is likely to provide a more profitable outlet for the rural landholder, without subsidies or credit for seed, fertilizer and machinery, and without a marketing system at the end of the process."

Because African agriculture after the onset of colonialism was and still is unprofitable, men were forced to migrate to urban areas, the mines, or commercial farming areas to work for wages to pay their taxes, invest in agriculture, supplement their agricultural incomes, and even purchase food. Dependency on migrant labor, especially at a household level, was consequently high (Taylor 1986). The loss of men to migration was taken as one of the factors contributing to low agricultural productivity (Schapera 1947, 162–67). Indeed, a recent study of profiles of rural households established that migrants' households dependent on hired labor were severely disadvantaged (Van der Wees 1981). The weight of evidence, however, also shows that migrant labor earnings are a crucial source of agricultural investments (Van der Wees 1981; Taylor 1986). Not surprisingly, those who remain in the rural areas work their land grudgingly, with one eye on the urban scene, and are prepared to abandon farming as soon as they secure a paid job. The irony of the situation is that urban wages were kept low based on the assumption that Africans were also engaged in agriculture. Riddell has observed that the wages paid to this generally unskilled labor force were geared to the subsistence needs of individual migrants, no allowance being made for family members or postemployment needs, and township facilities were rudimentary (Riddell 1980, 7). In Southern Rhodesia in 1977, 85 percent of all African urban employees received cash wages "below the austerely-calculated family poverty line and over 50% received wages less than half their minimum income requirements" (Riddell 1980, 7). All this emphasizes that outside their reserves Africans were (and still are in South Africa) regarded as temporary sojourners, and various forms of influx control measures such as the obnoxious pass laws have been introduced to control their movements into and out of the European areas.

As in America and other semiarid parts of the world, colonial rule not only limited the amount of land available to indigenous people but also put a brake on the movement that characterized the exploitation of natural resources in earlier times. The African areas have been experiencing growing population pressure for decades, and more than twenty years ago Monica Wilson noted that the demand for land among the Xhosa had increased to a stage where even though the commons were being converted into fields, there was an increasing landlessness (Wilson 1963, 382–83). In Southern Rhodesia in the late 1970s, available land in African areas could carry a maximum of 275,000 farming units and provide a modest income given the current levels of capitalization. But this land was holding about three times as many units, plots were too small to maintain families, and landlessness had increased to the point where 40 percent of rural men aged sixteen to thirty had no land (Riddell 1980, 3). The consequence was that the land was degraded because of overcultivation and overstocking; it was estimated that over seventeen times too much land was being cultivated—land that had been taken away from grazing land. Half of the grazing land either was completely bare or was heavily overgrazed. Soil erosion was rampant. Each year the land became less and less productive. Again, as Monica Wilson noted with respect to the Xhosa: "With the grazing now available, a quarter to a third of the cattle die in every drought. Though the cattle are fewer, now, than the men, they, together with the sheep and goats grazed, are nearly twice too many for the pasturage available" (Wilson 1963, 377–78).

Overgrazing is a general problem in the region as a whole. A recent paper, for example, observed that Swaziland has the highest cattle density per hectare in Africa and that "over the last twenty years livestock numbers have multiplied rapidly and the pressure on land has become increasingly severe, and is aggravated by the maldistribution of stock" (Fowler 1980). The environment is everywhere under severe stress. Devitt has observed that the Bakgalagadi of Botswana have allowed their flocks to increase on decreasing land and have not adopted any intensive range management practices: "As a result, the capital fund of energy is being used up faster than it is being replenished, and it accounts for the periodic, and apparently increasing frequent, collapse of the output flow. This is known as 'drought' and is largely a manifestation of a decreased stress tolerance in the local ecosystem" (Devitt 1977, 195).

Restricting the movement of groups has stabilized their settlement, encouraged further by the use of permanent materials in building houses and the provision of physical and social infrastructure that cannot be abandoned (Silitshena 1983, chap. 5). The areas near the villages are consequently denuded of vegetation except for some bushes (Axelsen 1977). In parts of Botswana, people are cultivating lands far from the villages as the lands nearby become exhausted, and they either purchase firewood or fetch it at increased distances (Silitshena 1983, chap. 5). In general there is no more unused land to shift to, and everywhere in the region the overcultivated soils are yielding diminished returns.

Other related and reinforcing factors in addition to land shortage have brought about this situation. The African population of Zimbabwe, for example, was estimated at 500,000 in 1890 (Kay 1970, 79) and has increased to 7

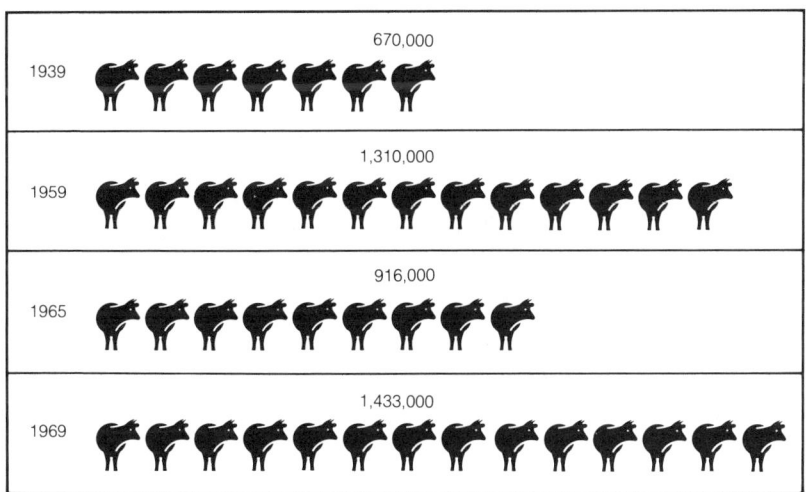

Figure 5.1 Growth of cattle population in Botswana.

million. One of the major causes of the rapid population growth (3–4 percent per annum) is the provision of medical facilities, but unfortunately, urban employment has not grown as fast as the population has, and the overcrowded rural areas have been relied on to carry increasing numbers of people. Veterinary services have had the same effect on livestock populations. Although there is periodic culling of livestock by drought in southern Africa, the trend is toward increasing numbers (figs. 5.1 and 5.2). Cattle continue to be an attractive form of investment not only because of traditional attitudes toward livestock but for sound rational reasons. Koijman (1978) has reviewed the role of cattle among the Tswana:

> The conclusion that can be drawn from the above is that for the vast majority of population cattle are not just "prestige objects needed for social purposes" as government officials have often asserted. The solid purpose if not of having directly increasing consumption is at least of having the assurance that should hardships or social emergencies befall the owner, he will be able to cope without being reduced to poverty. This partly explains the acquisition of as many cattle as possible. To the average farmer a large number of cattle seems to be the only way to survive a severe drought. (Koijman 1978, 181–82)

In any case, livestock have greater investment potential in some countries

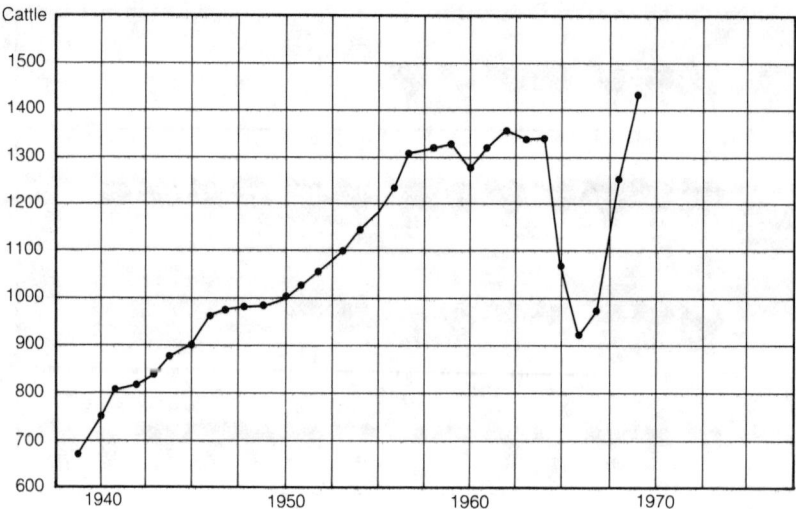

Figure 5.2 Growth of cattle population in Botswana.

than any other alternative open to people in the rural areas (e.g., Christopher 1982). As the human and cattle population grow apace, new technology, particularly the plow, further degrades the land. The near-universal use of the plow permits cultivation of more land per head than the hoe and therefore has the potential of exposing much larger areas to soil erosion (Kay 1970, 80–81). Accompanying the adoption of the plow have been changes in dietary preferences, so that the more environmentally demanding maize (corn) has everywhere supplanted the more tolerant sorghum and millet. Returns from maize are lower and less reliable, and to counter this, larger areas have been cultivated. Since there is no security in urban areas, everybody has had to maintain various kinship bonds in order to have access to land, and this has resulted in serious fragmentation of holdings.

Some efforts have been made to arrest the problem, especially in Southern Rhodesia. In South Africa, only after 1955 were major efforts made to tackle the problems arising from land degradation and conflicting land uses (Christopher 1982, 61–64). In Botswana the solution to overstocking in the densely populated eastern belt was to open up the fragile Kalahari to grazing, after digging boreholes. And in Lesotho the problem of acute soil erosion has still to be solved. Indeed, the three former protectorates are currently in various stages of developing solutions to these problems, such as changes in land tenure.[5] Southern Rhodesia made the most extensive efforts. In 1926 the white government attempted to solve the problems by reorganizing land uses in the African reserves by consolidating fragmented holdings and grazing land into blocks separated by residential zones. In this way damage to crops would be avoided and the two forms of activity harmonized (Kay 1970, 80–84). A South African variant of this solution has regrouped arable lands so that those on steep or eroded slopes were abandoned, the villages and cultivated land were fenced, and the rest of the land was used for grazing (map 5:4).

In 1941 the white government in Southern Rhodesia passed the Natural Resources Act, but without machinery to enforce it (Kay 1970, 80–84). In 1944 compulsory destocking was instituted in some forty reserves that were considered overstocked, an exercise Kay described as follows: "The owners could sell or slaughter their excess stock. No price-support programme was launched, and many Africans (probably with justification) felt that prices paid were unrealistically low. The fact that the purchasers were predominantly Europeans, mostly farmers or butchers, added to African resentment. Government arguments that such actions were to safeguard the land from the Africans and the tribesmen from themselves carried no weight in the African areas, and compulsory conservation measures could hardly have made a less auspicious beginning" (Kay 1970, 86).

Finally, in 1955 the government of Southern Rhodesia implemented a Land Husbandry Act whose aim was to deal with the "three basic needs of native agriculture: conservation, stability of tenure, and improving income out of which better practice can be funded." The objects of the act were:

(a) to provide for a reasonable standard of good husbandry and for the protection of natural resources by all Africans using the land and to enforce the maintenance of such a standard;

(b) to limit the number of stock in any area to its carrying capacity and, as far as practicable, to relate stock holding to arable landholding as a means of improving farming practice;

(c) to allocate individual rights in the arable land and in the communal grazing lands as far as possible in terms of economic units; and where this is not possible because of overpopulation, to prevent further fragmentation and to provide for the aggregation of fragmentary holdings into economic units;

Map 5.4 Village regrouping in Lebowa, South Africa.

(d) to provide for the setting aside of land for towns and business centers in the reserves. (Government of Southern Rhodesia 1955, 4)

The government experienced problems in implementing the act and eventually abandoned its attempts in 1962. One of its major problems was the shortage of land. For example, in 1959 the 102,000 families entitled to land and holdings that could not be allocated represented 30 percent of the cultivators and a population of over 500,000, and the government had to look around for empty spaces to resettle them. The expected boom in the industrial sector did not come about; "instead a recession took place which left many with the unhappy sense of insecurity of not knowing where to derive their immediate future livelihood" (Yudelman 1964), and African opposition to the act became intense, no doubt fueled by the suspicion that this was another device by the white government to take away the remaining African lands. The Land Husbandry Act has been criticized for being concerned with conservation and not with "people and their problems." As Yudelman has remarked, from the point of the soil scientist it was a very sound proposition (1964, 126), but it did not tackle the problems of insufficient land, lack of credit, poor access to markets, and low prices; stock limitation for many households with small herds would have been economically disastrous.

Africans have consistently refused to cooperate with white governments, because to them the problems of land degradation have been caused by the white men's taking a lion's share of the land (Wilson 1963, 387), and that some of the European land was lying idle has not helped the situation. At one time as much as 40 percent of European land in Southern Rhodesia fell into this category (Riddell 1980, 5).

In general, colonialism has destroyed or nearly destroyed the mixed pastoral/agriculturalist system that the southern African societies had developed for semiarid grazing areas, degrading the land and the soil and keeping people, now confined to reserves, from moving as their ancestors once moved to make a good use of the plant ecology of the region or from planting appropriate crops such as millet or sorghum rather than maize. The changes have also had social effects—for example, among the Tswana, undermining the power of the chief and encouraging a commercialization of agriculture that destroyed the nucleated settlement. The net effect of destroying the traditional agriculture and creating the reserves has been as devastating for the environment, especially soil and plant ecology, and for African societies as the development of a parallel system has been for the environment of the American Great Plains and that region's Indian societies.

NOTES

1. Some of these countries have changed their names since they achieved their independence: Basutoland to Lesotho (1966), Bechuanaland to Botswana (1966), and Southern Rhodesia to Zimbabwe (1980)

2. It has been argued that reviving these systems might provide an answer to the problem of livestock management and overgrazing in communal areas (Gulbrandsen 1980).

3. One chief at the time is quoted as complaining that "his people are not afraid to tell him that he is no longer able to do what he likes as the white people are now masters" (Dachs 1975, 16).

4. In Southern Rhodesia, African-produced maize fetched lower prices than that produced by whites. The reasons given were that prices should be fixed according to need; that African-produced maize was poor in quality; that it being essential to restore the fertility of European maize farms, European maize growers necessarily required a higher price than the Africans; that the standards of living were different; and that the African received free land, whereas the European purchased his.

5. Botswana launched the National Policy on Tribal Grazing Land in 1975; one of its major aims was to conserve the range. It tried to achieve this by modifying land tenure in communal areas. It was argued (Parson 1981, 236–55) that the measure might not achieve its main aims and would help only the large farmers. That the policy aims are not being achieved is now being officially acknowledged (Presidential Commission on Land Tenure 1983, 13). Lesotho recently introduced a land act that, among other things, takes away the functions of allocating land from the "corrupt" and "powerful" chiefs and leaves it to newly created institutions to ensure equitable distribution of land (Mosaase 1982).

REFERENCES

Arrighi, G. 1967. *The political economy of Rhodesia*. The Hague: Mouton.

———. 1970. Labour supplies in historical perspectives: A study of the proletarianization of the African peasantry in Rhodesia. *Journal of Development Studies* 6(3): 197–234.

Axelsen, B. 1977. *Land use and ecological problems in two villages in Botswana*. University of Oslo: Department of Geography.

Beach, David. 1977. The Shona economy: Branches of production. In *The roots of rural poverty in central and southern Africa*, ed. R. Palmer and N. Parsons. London: Heinemann.

Christopher, A. J. 1982. *South Africa*. London: Longman.

Cooke, H. J., and R. M. K. Silitshena. 1986. *Botswana: An environmental profile*. Report prepared for the United Nations Environmental Programme, Nairobi.

Dachs, A., ed. 1975. *Papers of John MacKenzie*. Johannesburg: Wits University Press.

Devitt, Paul. 1977. Coping with drought in the Kalahari. In *Drought in Africa*, ed. David Dalby et al. London: International African Institute.

Ettinger, S. 1972. South Africa's weight restrictions on cattle exports from Bechuanaland, 1924–1941. *Botswana Notes and Records* 4:21–29.

Floyd, B. N. 1962. *Changing patterns of African land use in Southern Rhodesia*. Lusaka: Rhodes-Livingstone Institute.

Fortmann, Louise. 1981. *Women's agriculture in a cattle economy*. Gaborone: Ministry of Agriculture, Rural Sociology Unit.

Fosbrooke, H. A. 1971. Land and population. *Botswana Notes and Records* 3:172–87.

Fowler, M. H. 1980. *The approaching crisis: Population, land and agricultural production in Swaziland*. Paper presented at the National Symposium on Population and Development, University of Swaziland, Institute of Economic and Social Studies.

Government of Southern Rhodesia. 1955. *What the Native Husbandry Act means to the rural African and to Southern Rhodesia*. Salisbury (Harare): Government Printer.

Gulbrandsen, Ørnulf. 1980. *Agro-pastoral production and communal land use: A socio-economic study of the Bangwaketse*. Gaborone: Ministry of Agriculture, Rural Sociology Unit.

Hitchcock, R. 1979. The traditional response to drought in Botswana. In *Proceedings of the Symposium on Drought in Botswana*, ed. M. T. Hinchey. Gaborone: Botswana Society.

———. 1982. Competition and conflict: Peasants, pastoralists, hunter-gatherers. In *Land reform in the making*, ed. R. P. Werbner. London: Rex Collings.

Houghton, D. Hobart. 1976. *The South African economy*. 4th ed. Cape Town: Oxford University Press.

Johnson, R. W. M. 1968. *The economics of African agriculture in Southern Rhodesia: A study in resource use*. Ph.D. thesis, University of London.

Kay, George. 1970. *Rhodesia: A human geography*. London: University of London Press.

Koijman, K. 1978. *Social and economic change in a Tswana village*. Leiden: African Studies Centre.

Kuper, Adam. 1975. The social structure of the Sotho-speaking peoples of southern Africa: Parts I and II. *Africa*. 45:67–81, 139–49.

Kuper, Hilda, et al. 1954. *The Shona and Ndebele of Southern Rhodesia*. London: International African Institute.

Lee, Richard B., and Irven DeVore. 1976. *Kalahari hunter-gatherers: Studies of the !Kung San and their neighbors*. Cambridge: Harvard University Press.

Legassick, Martin. 1977. Gold, agriculture, and secondary industry in South Africa, 1885–1970: From periphery to sub-metripole as a forced labour system. In *The roots of rural poverty in central and southern Africa*, ed. Robin Palmer and Neil Parsons. London: Heinemann.

Lye, William F., and Colin Murray. 1980. *Transformations on the highveld: The Tswana and Southern Sotho*. Cape Town: David Philip.

Marks, Shula, and Anthony Atmore, eds. 1980. *Economy and society in pre-industrial South Africa*. London: Longman.

Mason, Philip. 1958. *The birth of a dilemma: The conquest and settlement of Rhodesia*. London: Oxford University Press.

Massey, D. 1978. A case of colonial collaboration: The hut tax and migrant labour. *Botswana Notes and Records* 10:95–98.

Mosaase, A. 1982. *Lesotho's land policy under the Land Act 1979 and its implications on agricultural sector*. Paper presented at the Workshop on Land Policy and Agricultural Development, University of Botswana, February 14–20.

Omer-Cooper, J. D. 1966. *The Zulu aftermath: A nineteenth-century revolution in Bantu Africa*. London: Longman.

Palmer, Robin, and Neil Parsons, eds. 1977. *The roots of rural poverty in central and southern Africa*. London: Heinemann.

Parson, Jack. 1981. Cattle, class and the state in rural Botswana. *Journal of Southern African Studies* 7(2): 236–55.

Parsons, Neil. 1977. The economic history of Khama's country in Botswana, 1844–1930. In *The roots of rural poverty in central and southern Africa*, ed. Robin Palmer and Neil Parsons. London: Heinemann.

———. 1982. *A new history of southern Africa*. London: Macmillan.

Pauw, B. A. 1960. Some changes in the social structure of the Tlhaping of the Taung reserve. *African Studies* 19(2): 49–76.

Presidential Commission on Land Tenure, Republic of Botswana. 1983. *Report of the*

Presidential Commission on Land Tenure. Gaborone: Government Printer.

Riddell, Roger. 1980. Zimbabwe's land problem: The central issue. In *From Rhodesia to Zimbabwe: Behind and beyond Lancaster House*, ed. W. H. Morris-Jones. London: Frank Cass.

Sansom, Basil. 1974. Traditional economic systems. In *The Bantu-speaking peoples of southern Africa*, ed. W. D. Hammond-Tooke. London: Routledge and Kegan Paul.

Schapera, I. 1943. *Native land tenure in the Bechuanaland Protectorate.* Cape Town: Lovedale Press.

Schapera, I. 1947. *Migrant labour and tribal life: A study of conditions in the Bechuanaland Protectorate.* London: Oxford University Press.

————. 1953. *The Tswana.* London: International African Institute.

————. 1955. *A handbook of Tswana law and custom.* 2d ed. London: Frank Cass.

Shillington, K. 1985. *The colonization of the Southern Tswana, 1870–1900.* Braamfontein: Ravan Press.

Silitshena, R. M. K. 1972. The development of African agriculture in Rhodesia, 1900–1970: A study in economic stagnation. M.A. thesis, University of London.

————. 1983. *Intra-rural migration and settlement changes in Botswana.* Leiden: African Studies Centre.

————. 1985. *Social science research on drought in Botswana, Lesotho and Swaziland.* Nairobi: International Development Research Centre.

Taylor, John. 1986. Some consequences of recent reductions in mine labour recruitment in Botswana. *Geography* 71(1): 34–46.

Thompson, Leonard. 1969. Co-operation and conflict: The Zulu kingdom and Natal. In *The Oxford history of South Africa*, ed. Monica Wilson and Leonard Thompson. Oxford: Clarendon Press.

Tlou, T., and A. Campbell. 1984. *History of Botswana.* Gaborone: Macmillan.

Tyson, P. D. 1979. Southern African rainfall: Past, present and future. In *Proceedings of the Symposium on Drought in Botswana*, ed. M. T. Hinchey. Gaborone: Botswana Society.

Van der Wees, C. M. 1981. *The impact of migration on arable agriculture: A study in eastern Botswana.* Amsterdam: Free University.

Werbner, Richard P. 1971. Local adaptation and the transformation of an imperial concession in north-eastern Botswana. *Africa* 41(1): 32–41.

Wilson, Monica. 1963. Effects on the Xhosa and Nyakyusa of scarcity of land. In *African agrarian systems*, ed. Daniel Biebuyck. London: International African Institute by Oxford University Press.

————. 1969a. The Nguni people. In *The Oxford history of South Africa*, ed. M.

Wilson and L. Thompson. Oxford: Oxford University Press.

————. 1969b. The Sotho, Venda and Tsonga. In *The Oxford history of South Africa*, eds. M. Wilson and L. Thompson. Oxford: Oxford University Press.

————. 1969c. The hunters and herders. In *The Oxford history of South Africa*, eds. M. Wilson and L. Thompson. Oxford: Oxford University Press.

Wilson, N. H. 1923. The development of native reserves. NADA I (December).

Yudelman, Montague. 1964. *Africans on the land*. Cambridge: Harvard University Press.

Chapter Six

Plains Indians and Australian Aborigines in the Twentieth Century

Peter Iverson

In writing of the ancient, one-eyed Kiowa woman Ko-sahn, N. Scott Momaday contended that for her "there was no distinction between the individual and the racial experience, even as there was none between the mythical and the historical. Both were realized for her," he observed, "in the one memory, and that was of the land." Momaday added: "This landscape, in which she had lived for a hundred years, was the common denominator of everything that she knew and would ever know—and her knowledge was profound. Her roots ran deep into the earth, and from those depths she drew strength enough to hold still against all the forces of chance and disorder. And she drew therefrom the sustenance of meaning and of mystery as well" (Momaday 1981, 166–67).

One hundred years ago, conventional wisdom was that the Plains Indians of the United States and the Aborigines of Australia were about to disappear. The original occupants of their lands, these peoples seemed doomed to extinction as separate, identifiable participants within their regions. The nonindigenous population grew rapidly, and cattle ranchers and others justified expropriation of native lands. The state and national governments, through their laws and the actions of their officials, left little doubt about their sympathies. One could predict confidently, well before the late twentieth century, that a thorough assimilation of the Plains Indians and the Aborigines would be achieved.

Conventional wisdom would be proved wrong, and the twentieth century would be characterized ultimately by survival and continuation (Iverson 1985b, introduction). Nonetheless, a demographic transition had occurred by the turn of the century. The 1901 census counted 67,000 Aborigines remaining in Australia, compared with perhaps 250,000 in 1788 when the country was established officially as a colonial entity (Harris 1979, 36). The overall population of Indians within the United States, of course, had declined considerably since the incursion of the Europeans. In the Plains, smallpox and war took their toll, though the population of Sioux, for example, expanded considerably in the eighteenth and nineteenth centuries (White 1978, 319–43; Lowie 1982, 9–12). But the prevailing image still depicted defeat and eventual disappearance. Restricted in their movements, the subject of widespread efforts toward assimilation, the people themselves could not have been optimistic about their future.

We usually think of the ghost dance in this context, and until rather recently we looked to the early twentieth century as a time of unrelieved despair for American Indians in general and Plains Indians in particular. Black Elk is often quoted in this context: "And I can see that something else died there in the bloody mud, and was buried in the blizzard. A people's dream died there. It was a beautiful dream. . . . And I, to whom so great a vision was given in my youth—you see me now a pitiful old man who has done nothing, for the nation's hoop is broken and scattered. There is no center any longer, and the sacred tree is dead" (Neihardt 1961, 276).

However, we tend to forget the postscript to the book. Neihardt and Black Elk journey to Harney Peak, where amid the clouds and the rain Black Elk issues a last appeal: "It may be that some little root of the sacred tree still lives. Nourish it then, that it may leaf and bloom and fill with singing birds. Hear me, not for myself, but for my people; I am old. Hear me that they may once more go back into the sacred hoop and find the good red road, the shielding tree!" Neihardt's concluding sentence informs us, "In a little while the sky was clear again" (Neihardt 1961, 276–80; Iverson 1984, 100–109).[1]

To appreciate the nature of Indian and Aboriginal resilience, one must review some of the most threatening obstacles the Plains Indians and Aborigines have had to confront. In addition, one must examine some of the most significant developments among these people that have contributed to their survival and continuation. Although there have been important differences in the historical and contemporary experiences of the two groups, it is useful to recognize critical commonalities.

The obstacles may be summed up readily, if perhaps too simply. The rest of the population has had severe prejudices about people different from themselves in appearance and seemingly in culture. That population, which has increased dramatically in size, has had other priorities for the lands and lives of the indigenous peoples. Local, state, and national governments have reflected the sentiments and the will of the majority. Through laws, court decisions, and other manifestations, the government has implemented policies and practices centrally influencing Plains Indian and Aboriginal life.

Some of the more distinct patterns of prejudice vary from one generation to the next, but there are core attitudes that remain—indeed are passed down, inherited, learned from the cradle. These moral judgments tend to be severe and unyielding, and there is little sign that they are softening much in areas where the indigenous population is more than a small fraction of the citizenry. Walk down Todd Street in Alice Springs or along Omaha in Rapid City, and one hears and sees the same thinking at work. Michael Heppell and Julian Wigley (1981) summarize the "prevailing white view" of Aborigines in Alice Springs: "'They' are dirty, lazy, live in squalor, receive generous government handouts (which are withheld from needy whites) which they spend irresponsibly, drink too much but cannot hold their drink and therefore brawl too frequently. 'They' also secretly lust after white women and, like animals, barely keep these desires below the surface." Heppell and Wigley are especially disturbed by the absence "of interest in and sympathy for the causes of the present conditions suffered by so many Aborigines." They note that "the enduring official and unofficial attitude has long been that Alice Springs should be purged of Aborigines" (Heppell and Wigley 1981, 40). Loretta Fowler's description of white attitudes in the bordertowns next to the Wind River reservation is almost identical (Fowler 1985, 187–217). Or to employ the analysis of Niels Winther Braroe about whites and Chippewas-Crees in a Canadian Plains community, Indians seem irresponsible, untrustworthy, and childishly impulsive; whites respond with certain kinds of recurrently applied sanctions, just as sanctions are applied or attempted in Alice Springs (Braroe 1975, 116).

Within rural Australia as within the rural Plains, the fates of Aborigines or Indians and whites have been interwoven from the beginning. The pastoral or cattle-ranching industry early on became vital in both regions. Long-term leasing emerged as a prevalent practice, though the contract in Australia would be made with the state or territorial government rather than through the federal government—a contrast mirroring heightened state and territorial authority within that nation (Stevens 1974). For Aborigines, working on pastoral stations permitted continuing residence within their home country. Their labor on the remote Northern Territory stations in the early twentieth century, in the judgment of Special Commissioner Baldwin Spencer, made "all the difference at the present time between working the station at a profit or a loss" (Donovan 1984, 20–21). However, that labor was compensated minimally, if really at all (Altman and Nieuwenhuysen 1979, 59–75). Similarly, sizable reductions in the Plains Indian estate through "agreements" pertaining to sale or lease permitted the white cattle rancher an opportunity to flourish. Although there were examples of successful Indian cattle ranching, support for such enterprises generally did not materialize for extended periods, even as federal officials and local whites bemoaned Indian laziness (Iverson 1986). In a recent study of the economic status of Aborigines, Jon C. Altman and John Nieuwenhuysen perceived the same dichotomy that has characterized American federal policy: "It is difficult to see how the policy of assimilation related to the location of most government settlements and missions in isolated regions with scarce natural resources and a harsh climate. It is also difficult to see how the fostering of Aboriginal reliance upon government welfare and transfer payments (and the absence of employment opportunities in European-style jobs, even where training for these was provided) could be reconciled with the assimilation policy" (Altman and Nieuwenhuysen 1979, 24).

In both countries policies toward the peoples have evolved through stages identified by some of the same phrases or sentiments; assimilation and self-determination have been talked about and at times encouraged. Yet the rights of both groups have been severely circumscribed. The right to vote, the right to equal educational opportunities, the right to live where one wants, the right to speak a language other than English, the right to believe in religions other than Christianity—all these and other rights have been argued about, fought for, and in some instances won. Government officials and social scientists talk about full-bloods and mixed-bloods in both countries, sometimes critique popular attitudes, perhaps urge alteration of existing procedures. Occasionally agents who live with the people long enough and more frequently than anthropologists and other academic observers press for better treatment and more complete understanding, though they risk being accused of being sentimentalists or separatists. Vested interests intent on winning indigenous souls or property quickly counterattack. Catherine H. Berndt remembered that when anthropologists in the 1930s, 1940s, and 1950s "pressed for the establishment and safeguarding of Aboriginal reserves, they were accused of supporting a policy of apartheid" (Berndt 1977, 402). Well-meaning white liberals have been fond of labeling reservations "concentration camps," a label rejected with increasing vehemence by reservation residents, who regard them as simply their homes (Giago 1984, 94–97).

Within the Plains, the movement to transform reservations from prisons to homes is of long standing. On the Cheyenne River reservation in central South Dakota, for example, this transition could be observed in its early forms before World War I. Because of concerted action on the part of Cheyenne River Sioux, their land base was not eliminated; despite the best efforts of Senator Robert Gamble and other would-be empire builders, the people held on to their land (Hoxie 1985, 55–75). The maintenance of a land base has been critical in regard to the functioning of an adaptive, changing culture. The reservation boundary is part of a working ethnic boundary, even though it may be perforated by the interests of outside forces (Barth 1969, 9–38).

Even given the efforts to maintain boundaries, to make often marginal land more productive in order to allow as many of the people as possible to stay rather than to migrate for economic reasons, Plains Indians live under what Ann Laquer Estin has termed "the long shadow" of *Lone Wolf v. Hitchcock*. This decision by the United States Supreme Court in 1903 meant that "all aspects of tribal political and property rights were now subject to radical changes at the whim of the legislative branch. At best, tribes might retain rights only so long as more powerful political pressure-groups had no interest in their Indian resources and no objection to their continued existence" (Estin 1984, 240). Although the courts have softened the blow to some degree during the course of the century, Congress clearly is in charge; if the courts occasionally will protect Indian rights, Congress rarely will (Estin 1984).[2]

In both Australia and the United States, the native peoples speak about sovereignty and look to judicial opinions to reinforce their rights. When it comes to the day-to-day jousting, of course, it is with elected and appointed government officials that they have to contend. This is not to suggest the justices have been especially more enlightened than legislators, but only to emphasize a continuing truth from the days of Andrew Jackson and John Marshall, when a president's power mattered more ultimately than a justice's decision. In Australia, with the heightened powers of local, state, and territorial governments, the point is even more important. Queensland is perhaps the most notorious example, spurred by the rigid views of its premier for many years, Johannes Bjelke-Petersen. Here Aborigines consistently have been limited in their rights because of a hostile government at the regional level (Nettheim 1981).

Although in more recent years Aborigines and Plains Indians have become a less rural populace, sizable numbers still reside within more remote portions of their regions. These lands are frequently important for traditional cultural reasons, including many sacred sites. The economic development of such land has always posed a challenge, with ranching and mineral exploitation emerging as primary modes. Mining has been promoted not only by local individuals, who seize upon it as a means of encouraging development in general, but also, to be sure, by citizens far removed from the site of proposed mining who stand to gain by eventual domestic consumption of mineral products. In more recent years, multinational interests have also become involved in this scenario.

The growing assertion of Plains Indian and Aboriginal land rights has conflicted directly with the perceived needs of state and territorial economies. Mining officials and government figures frequently have argued against the efforts by Indians and Aborigines to control the extent of mineral exploration and exploitation.[3] The response by the people themselves understandably has often been bitter. Regardless of internal debates and doubts about the disruption posed by mining, the people often have been disappointed by the financial returns. The effect upon employment has generally been less than anticipated, and the sums added to various treasuries have frequently proved much reduced from initial optimistic projections. The pace of negotiation, the extent of disruption, and the long-term repercussions of decisions made all pose serious dilemmas for the responsible parties.[4]

Such problems may be presented in both negative and positive terms. Social scientists and investigative journalists often focus on the dysfunctional aspects of contemporary developments. On the other hand, the very fact that we are talking about land rights indicates change in the perspective of a generation ago; in the words of Nicolas Peterson, in 1966 in Australia there simply were no Aboriginal land rights, but now there are (Peterson 1981, 1). There have not always been land councils in Australia, nor have Indians in the United States always had the right to tax Indian and non-Indian businesses on reservations without first receiving the approval of the federal government.[5] If we acknowledge continuing and severe liabilities, what are some of the elements within Aboriginal and Plains Indian life that have contributed toward survival and at time improvement of people's lives?

First of all, we should note that while activities by Plains Indian and Aboriginal leaders have been more publicized within the past generation, their efforts are built upon a foundation established gradually during this century. Concern for sacred sites, efforts to improve the well-being of the people, and resistance to erosion of existing rights all represent a legacy of earlier endeavors by their predecessors. As the study of Plains Indian history in this century matures, we are learning more about these relatively unsung pioneers for Indian rights; the same is true for Aboriginal history. The Society of American Indians and the Australian Aborigines' League, Zitkala-Ša and William Cooper begin to emerge from obscurity (Hertzberg 1971; Welch 1985; Jones and Hill-Burnett 1982). The post–World War II period, for example, generally referred to in the United States as the termination era, actually saw as well the growth of tribal and pantribal developments in a variety of areas, including economy, organization, education, law, and politics. The same period also was a critical time for pan-Aboriginal movements (Iverson 1985a, 163–73).

Although tribal, clan, and linguistic distinctions have remained important among Indians and Aborigines, such differentiation has largely been ignored by other Americans and Australians. Indians are seen as Indians, in other words, rather than as members of tribes or other communities. The growing urbanization of the Indian population, particularly since the Second World War, has naturally contributed to the growth of pan-Indian organizations and another level of Indian identity beyond that of tribe. Marriage outside the tribe, of course, has also increased. Through this century, as educational and travel opportunities have been presented, more Indians have acquired a common language—English—and used it to improve communication among themselves. Specific problems faced in the city also have heightened many Plains Indians' sense of urgency and at times militancy, with organizations such as the American Indian Movement coming out of that urban experience (Witt 1970; Thomas 1970; Fixico 1986).

For some time, early pan-Aboriginal efforts were limited by constraints like those that had faced Plains Indians, including difficulty of communication, limitations on the movements of the people, and a small urban population. After World War II, Aborigines were no longer banned, as they had been, from residing in cities such as Perth. In addition, because of increased mechanization in farming and ranching, their manual labor often proved less valued, and many people were forced into the city seeking economic alternatives. It was in the Australian towns and cities that pan-Aboriginal movements would really begin to evolve (Jones and Hill-Burnett 1982, 220–21; Howard 1982, 6–7).

More people relocated to towns and cities, then, but that did not mean they had removed themselves entirely from what many referred to as home or their home country. Urban residents traveled back and forth from town to country, and even while living in a town such as Alice Springs, Aborigines maintain ties with kinsmen in traditional country, and their daily activities are seemingly comparable. People travel to the country for certain ceremonies. Those who camp within the town generally organize their camps based upon geographical and linguistic affiliations, and the camps are usually established in the part of town in the direction of their own traditional country (Heppell and Wigley 1981, 51–56).

Poor housing, discrimination, and lack of educational opportunities within the towns and cities also provoke and promote activism.[6] In attempting to remedy various situations, Plains Indians recognized the potential importance of legal counsel, and the effect of those legal services did not go unnoticed in Australia. When the Labour party came to power in Australia in 1972, it pledged to improve Aboriginal access to legal representation. In both countries, attorneys have come to play roles beyond what federal authorities initially had anticipated and preferred. They have been vital forces in promoting indigenous rights, including questions surrounding land use (Eggleston 1977; Dobyns 1970). The issue of land tenure may be even more complicated in Australia, if that is possible, than in the United States; inevitably, anthropologists have also become centrally involved in various frays. One can only agree with Kenneth Maddock in his review of the perils of legal anthropology: "The confusion will be greatest where the anthropologist is trying to make sense of inside realities by accommodating them within an outside conceptual framework the terms and structure of which he does not properly understand" (Maddock 1981, 100). Claims cases in the Plains have been subject to much the same confusion and subsequent dissatisfaction.

Despite the inherent and inevitable confusion and dissatisfaction involving land claims and land rights, it is here that the vitality of modern Aboriginal and Plains Indian life is centered. It is from the land that the people draw, as Momaday (1981) suggests, the sustenance of meaning and mystery. The problems of rural America and rural Australia encompass more than the descendants of their original inhabitants; farmers, ranchers, owners of small businesses, and others, regardless of heritage, recognize in the land and in the overall environment of rural life something precious, something that has been handed down, something one lets go of only with the utmost reluctance. With the bewildering array of dilemmas confronting rural users of land, it is possible for a white rancher in Montana, for example, to appreciate a bond with Indians of the past and present; the possibility of being forced off the land can lead to the perception of commonality (Beer 1985, 62–64; Steiner 1985). It is more likely to lead to hostility and confrontation, as entrenched interests try to stave off significant change or even extinction. In Australia, thus, the efforts by Aborigines to force change has led to one struggle after another with the pastoral industry. The mining industry's interest in limiting Aboriginal control merely compounds the issue.

Although the movement for Aboriginal land rights enjoys a long history, a new chapter was written in August 1966 when the Gurintji went on strike at Lord Vestey's Wave Hill station in the Northern Territory. Two issues lay at the heart of the dispute: adequate wages and a desire for the return of traditional lands. The strike persisted for a decade and, together with other developments, helped galvanize popular sentiment, particularly in regions where Aborigines formed a much smaller percentage of the Australian population. Concerned about mining leases on their traditional lands in northeastern Arnhem Land, Yolngu clans at Yirrkala in December 1968 sought relief through the supreme court of the territory. The court ruled that the Aborigines did not at present have such land rights as could be recognized legally within the country (Howie 1981, 28; Layton 1985, 148–49). Mr. Justice Blackburn's decision was not entirely unsympathetic to Aboriginal interests. Although the Aborigines had a legal system, their form of communal title under the Australian system did not legally grant them the rights of a proprietor. In reviewing the ruling, Robert Layton states the Blackburn "recognized three aspects of proprietary right, the right to use and enjoy, the right to exclude others—which the plaintiffs had claimed only with respect to sacred sites, not land in general—and the right to alienate; a right deliberately repudiated by the plaintiffs" (Layton 1985, 149).

In the wake of the decision, the Labour government in Canberra appointed the Aborigines' attorney, Mr. Justice Woodward, Aboriginal land commissioner in February 1973. On record in support of legislation to provide land rights for the Aborigines—"not just because their case is beyond argument," said Prime Minister Gough Whitlam, "but because all of us as Australians are diminished while the Aborigines are denied their rightful place in this nation"—Labour charged Woodward with the task of locating "the appropriate means to recognize and establish the traditional rights and interests of the Aborigines in and in relation to land, and to satisfy in other ways the reasonable aspirations of the Aborigines to rights in or in relation to land" (Howie 1981, 28–29).

In Woodward's reports in July 1973 and May 1974 he recommended, among other things, transfer of Aboriginal reserves and missions with some additional land to full Aboriginal ownership, formation of Land councils, payment of mining royalties, purchase of cattle stations, and leasing of additional land. The eventual act gained approval over intense opposition from non-Aboriginal interests in the territory. The opposition did gain some victories, including restricting claims to those that could be lodged on traditional grounds rather than on the basis of need and empowering the Northern Territory Legislative Assembly to deal with several crucial concerns. These concerns included "protection of sacred sites, the control of entry to Aboriginal lands and adjacent waters," and "rights in relation to pastoral properties." Significant additional lands that Woodward had recommended be transferred to Aboriginal land trusts also were left out of the act (Howie 1981, 29–31).

Though weakened, the act far surpassed any piece of legislation that had gained approval up to that time in Australia, and its provisions naturally encouraged various interpretations. At the heart of the controversy, to be sure, would be claims relating to tradition and counterclaims relating to benefits that would accrue to mining companies, pastoralists, and the government of the territory (Layton 1985, 150–51). After the passage of the Aboriginal Lands Rights Act in 1976, the commonwealth minister for Aboriginal affairs, Ian Viner, wrote a series of thirteen articles for the *Northern Territory News* in which he attempted to show how the act would affect all who lived in the territory. Viner argued that the presentation to Aboriginal people of legal title to their own land "should mark the beginning of a new era of unity for all the people in the Northern Territory."[7]

The decade that has followed has not witnessed a fulfillment of Viner's vision. The outstation movement accelerated, whereby Aborigines occupied land they claimed through traditional rights, often with the purpose of protecting sacred sites.[8] Land councils established within the territory vigorously pressed for title to existing reserves and to vacant Crown land. Even though pastoral leases were almost entirely exempt from the act, and even though established mining leases were to be honored, a severe backlash greeted efforts to put the act into effect.[9] Where the act initially might have been perceived as a beginning, it is now often viewed by Aborigines more as a landmark that must be preserved and honored. Aboriginal criticism has heightened in the wake of the recent setbacks. In the September 1985 issue of *Land Rights News* Pat Dodson pilloried Prime Minister Robert Hawke for his Labor government's "about-face" on Aboriginal land rights and charged that the pastoral industry had not provided "justice for the people who built" that economic enterprise. With the Australian bicentennial of 1988 fast approaching, Dodson suggested that "if they really want to add some credibility to the celebrations, they should be sitting down now and negotiating an agreement, a treaty between Aboriginal people and themselves. That's the only honest thing they could do." On the advent of the bicentennial in the United States, Vine Deloria, Jr., in *Behind the Trail of Broken Treaties* (1985), made a similar plea for reinstitution of the treaty-making process.[10]

Today in the United States, Indian critics are equally severe in their pessimism and their demands for justice. Indian land rights seem tenuous. Attorneys Robert T. Coulter (an enrolled member of the Citizens Band of Potawatomi) and Steven N. Tullberg bluntly charge that "United States law and the courts have substantially failed to protect Indian lands from arbitrary confiscations, partitioning, bureaucratic control, and treaty violations." Their list is indeed staggering: (1) Indian land lacks constitutional protection from arbitrary seizure by the United States; (2) there is no protection against federal confiscation of "aboriginal title" or "Indian title" land (perhaps the majority of Indian land); (3) Congress may unilaterally abrogate Indian treaties; (4) the federal government says it has trust title to all Indian lands outside the original thirteen colonies; and (5) the law and the courts offer little hope against termination of Indian governments and tribes if the political climate would permit such action (Coulter and Tullberg 1984, 87–88).

Coulter and Tullberg (1984, 208) correctly emphasize that spiritual attachment to the land is central to the debate over Indian land rights. "The religious character of the land and the fundamentally different concept of land tenure of many Indian people has played a determining part in many of the unresolved conflicts," they observe, which "partly explains the phenomenal tenacity of Indian peoples in their land claims," be they of the Black Hills or elsewhere. Altman and Nieuwenhuysen (1979, 78) echo this sentiment of the Aborigines: "The special relationship which Aboriginal people have with the land seems best described by the word 'spiritual.'" Ultimately, we must remember that it is the spiritual component of the land rights battles in both countries that gave these confrontations their particular quality.

The controversies will continue. Still, the twentieth century has shown on balance a remarkable resilience on the part of the Plains Indians and the Aborigines. We know that they are far from vanishing peoples. Indeed, the populations of both groups continue to grow rapidly, with the most recent censuses recording striking demographic increases. And we should know that as always has been so, that the land is at the heart of it all (Iverson 1989). When one walks around or to the top of a sacred mountain in Plains Indian or Aboriginal country, be it what we would call Bear Butte or Nowah'wus, Ayers Rock or Uluru, one may gain that sense of power that the land commands. One may appreciate more fully that the people and the land will endure. Here it was, one remembers, that the people of long ago learned a lesson from culture heroes such as Sweet Medicine, who left their mark in an ancient ceremonial painting. Here it is that, as Peter Powell has written about the Cheyennes' holy mountain, one may gain strength and reassurance, "even in the strange new days that lie ahead" (Powell 1985, 262). The land remains; the people do too.

NOTES

1. For an extremely insightful view of the two men, see DeMallie (1984, 1–74).

2. See also the discussion of *Lone Wolf*'s significance in Hoxie (1984, 154–57, 164– 66, 170). A thorough analysis of the specific background may be found in Hagan (1976).

3. An informative overview of the land rights issues in the different areas of Australia is contained in Peterson (1981). For an example of the complex issues pertaining to control of mineral resources in the Plains, see Ambler (1984, 11).

4. The *Land Rights News* published by the Central Land Council in Alice Springs provides continuing coverage of controversies surrounding mining and pastoral issues. See, for example, "Land Rights: The Story So Far," in the September 1985 edition. For a consideration of Plains Indians and mineral development, see Donald Fixico, "Tribal Leaders and the Demand for Natural Energy Resources on Reservation Lands" (Iverson 1985b, 219–36).

5. The Navajo Nation in the United States has been among the Indian tribes most vigorous in pressing for extension of tribal authority. The case referred to here involved Kerr-McGee Corporation's effort over seven years to override the 3 percent tribal tax on the value of leasehold interests and the 5 percent tribal tax on the sale of property produced or extracted within the Navajo Nation. The United States Supreme Court ruled unanimously in favor of the Navajos on April 16, 1985. According to an article in the newsletter of the Indian Rights Association, *Indian Truth* (no. 263, June 1985), all American Indian tribes stand to benefit from the Court's decision, regardless of whether they are organized under the Indian Reorganization Act.

6. In both Australia and the United States, control of local education for their children has emerged as a priority. Plains Indian efforts have extended to the community college level, with most reservation communities now having such an institution. Public and contract schools offer new opportunities for Indian students. In Alice Springs, the Yipirinya School opened in 1979 as an Aboriginal community school controlled by an autonomous Aboriginal council. Founded in 1969 under church auspices, the Institute for Aboriginal Development is now directed by Aborigines and offers training in leadership, language, literacy, and community development. Information about these institutions is from literature obtained during my visit to Alice Springs in the summer of 1983.

7. Ian Viner's series of articles written in the latter half of 1978 were subsequently

published by the Department of Aboriginal Affairs through the Australian Government Publishing Service in 1979 in a twenty-four-page booklet entitled "Aboriginal Land Rights in the Northern Territory: What It Means and How It Will Work."

8. See, for example, the article on the Simpson desert outstation in *Land Rights News*, Central Australia edition, Autumn 1983.

9. the Central Land Council labeled 1984 "the year of the backlash against the movement for Aboriginal Land Rights in Australia." A recent publication printed in Alice Springs, "Still Waiting for My Country," quotes Davey Hayes: "We've done so much for the stations. The whitefellows got rich and we did all the work. . . . We worked hard but we got nothing. We've been pushed off our country. We've got to go back there."

10. Vine Deloria, Jr., originally made the point in *Behind the Trail of Broken Treaties: An Indian Declaration of Independence*, originally published in 1974. The book was recently reprinted (Austin: University of Texas, 1985). Dodson and other Aboriginal leaders have increasingly used international forums to learn about and discuss international indigenous peoples' rights.

REFERENCES

Altman, Jon C., and John Nieuwenhuysen. 1979. *The economic status of Australian Aborigines*. Cambridge: Cambridge University Press.

Ambler, Marjane. 1984. Indians will again take control of the coal they own. *High Country News* 16 (December): 11.

Barth, Fredrik, ed. 1969. *Ethnic groups and boundaries: The social organization of cultural difference*. Boston: Little, Brown.

Beer, Ralph. 1985. Holding to the land: A rancher's sorrow. *Harper's* 271 (September): 1624.

Berndt, Catherine H. 1977. Out of the frying pan . . . ? or, Back to square one? In *Aborigines and change: Australia in the 70's*, ed. R. M. Berndt. Social Anthropology Series 11. Canberra: Australian Institute of Aboriginal Studies.

Braroe, Niels Winther. 1975. *Indian and white: Self-image and interaction in a Canadian Plains community*. Stanford, Calif.: Stanford University Press.

Cadwalader, Sandra L., and Vine Deloria, Jr., eds. 1984. *The aggressions of civilization: Federal Indian policy since the 1880s*. Philadelphia: Temple University Press.

Coulter, Robert T., and Steven M. Tullberg. 1984. Indian land rights. In *The aggressions of civilization: Federal Indian policy since the 1880s*, ed. Sandra L. Cadwalader and Vine Deloria, Jr. Philadelphia: Temple University Press.

Deloria, Vine. 1985. *Behind the trail of broken treaties: An Indian declaration of independence*. Austin: University of Texas Press.

DeMallie, Raymond. 1984. Nicholas Black Elk and John G. Neihardt: An introduction. In *The sixth grandfather: Black Elk's teachings given to John G. Neihardt*, ed. R. DeMallie, Lincoln: University of Nebraska Press.

Dobyns, Henry F. 1970. Therapeutic experience of responsible democracy. In *The American Indian today*, ed. Stuart Levine and Nancy D. Lurie. Baltimore: Penguin Books.

Donovan, Peter F. 1984. *At the other end of Australia: The Commonwealth and the Northern Territory, 1911–1978*. St. Lucia: University of Queensland Press.

Eggleston, Elizabeth. 1977. Aboriginal legal services. In *Aborigines and change: Australia in the 1970's*, ed. R. M. Berndt. Social Anthropology Series 11. Canberra: Australian Institute of Aboriginal Studies.

Estin, Ann Laquer. 1984. *The long shadow: Lone Wolf v. Hitchcock*. In *The aggressions of civilization: Federal Indian policy since the 1880s*, ed. Sandra L. Cadwalader and Vine Deloria, Jr. Philadelphia: Temple University Press.

Fixico, Donald Lee. 1986. Termination and relocation: Federal Indian policy, 1945–1960. Albuquerque: University of New Mexico Press.

Fowler, Loretta. 1985. What they issue you: Political economy at Wind River. In *The Plains Indians of the twentieth century*, ed. Peter Iverson. Norman: University of Oklahoma Press.

Giago, Tim. 1984. Indian reservations: The only land we know. In *Notes from Indian country*, vol. 1, ed. Tim Giago. Pierre, S. Dak.: State Publishing Company.

Harris, Stewart. 1979. *"It's coming yet . . .": An Aboriginal treaty within Australia between Australians*. Canberra: Aboriginal Treaty Committee, 1979.

Heppell, Michael, and Julian Wigley. 1981. *Black out in Alice: A history of the establishment and development of town camps in Alice Springs*. Monograph 26. Canberra: Australian National University Development Studies Centre.

Hertzberg, Hazel W. 1971. *The search for an American Indian identity: Modern pan-Indian movements*. Syracuse: Syracuse University Press.

Howard, Michael C. 1982. *Aboriginal power in Australian society*. Honolulu: University of Hawaii Press.

Howie, Ross. 1981. Northern Territory. In *Aboriginal land rights: A handbook*, ed. Nicolas Peterson. Canberra: Australian Institute of Aboriginal Studies.

Hoxie, Frederick E. 1984. *A final promise: The campaign to assimilate the Indians, 1880–1920*. Lincoln: University of Nebraska Press.

———. 1985. From prison to homeland: The Cheyenne River reservation before

World War. I. In *The Plains Indians of the twentieth century*, ed. Peter Iverson. Norman: University of Oklahoma Press.

Iverson, Peter. 1984. Neihardt, Collier and the continuity of Indian life. In *A sender of words: Essays in memory of John G. Neihardt*, ed. Vine Deloria. Salt Lake City, Utah: Howe Brothers.

―――. 1985a. Building toward self-determination: Plains and southwest Indians in the 1940s and 1950s. *Western Historical Quarterly* 16(2): 163–73.

―――. 1985b. *The Plains Indians of the twentieth century*. Norman: University of Oklahoma Press.

―――. 1986. Cowboys, Indians and the modern West. *Arizona and the West* 28(2): 107–24.

―――. 1989. Cowboys and Indians, Stockmen and Aborigines: The Rural American West and The Northern Territory of Australia Since 1945. *Social Science Journal* 26(1): 1–14.

Jones, Delmos J., and Jacquetta Hill-Burnett. 1982. The political context of ethnogenesis: An Australian example. In *Aboriginal power in Australian society*, ed. Michael C. Howard. Honolulu: University of Hawaii Press.

Layton, Robert. 1985. Anthropology and the Australian Aboriginal Land Rights Act in northern Australia. In *Social anthropology and development policy*, ed. Ralph Grillo and Alan Rew. London: Tavistock.

Lowie, Robert H. 1982. *Indians of the Plains*. Lincoln: University of Nebraska Press.

Maddock, Kenneth. 1981. Warlpiri land tenure: A test case in legal anthropology. *Oceania* 52(2): 85–102.

Momaday, N. Scott. 1981. The man made of words. In *The remembered earth: An anthology of contemporary Native American literature*, ed. Geary Hobson. Albuquerque: University of New Mexico Press.

Neihardt, John G. 1961. *Black Elk speaks*. Lincoln: University of Nebraska Press.

Nettheim, Garth. 1981. *Victims of the law: Black Queenslanders today*. Sydney: George Allen and Unwin.

Peterson, Nicolas. 1981. *Aboriginal land rights: A handbook*. Canberra: Australian Institute of Aboriginal Studies.

Powell, Peter J. 1985. Power for new days. In *The Plains Indians of the twentieth century*, ed. Peter Iverson. Norman: University of Oklahoma Press.

Steiner, Stan. 1985. *The ranchers: A book of generations*. Norman: University of Oklahoma Press.

Stevens, Frank. 1974. *Aborigines in the Northern Territory cattle industry*. Aborigines in Australian Society II. Canberra: Australian National University Press.

Thomas, Robert K. 1970. Pan-Indianism. In *The American Indian today*, ed. Stuart

Levine and Nancy O. Lurie. Baltimore: Penguin Books.

Welch, Deborah S. 1985. Zitkala Ša: An American Indian leader. Ph.D. diss., University of Wyoming.

White, Richard. 1978. The winning of the West: The expansion of the western Sioux in the eighteenth and nineteenth centuries. *Journal of American History* 65:319–43.

Witt, Shirley Hill. 1970. Nationalistic trends among American Indians. In *The American Indian today*, ed. Stuart Levine and Nancy O. Lurie. Baltimore: Penguin Books.

Part Three

European and Indigenous Institutions

The removal of semiarid indigenous peoples from their traditional unappropriated lands required the creation of new institutions designed to accommodate the developmental interventions forced on them. The most obvious of these is the reserve discussed in the Introduction. In idea, the reserve appears both to mirror an old way and to announce "progress," but in fact, it usually does neither. Rather, it places its residents in a colonialist or neocolonialist "trust" relationship with the nation-state, severely limiting their traditional rights and territories and imposing new institutions on them. On the reserve, one will often find an education designed to force the "first people" of the area to adapt to Western ways, religious institutions designed to destroy their old religion, land-tenure corporations intended to do away with usufruct property concepts, and governance structures that substitute for traditional clan leadership.

This process of sedentarizing on reserves, introducing new property concepts, "educating," and replacing traditional leadership has taken place in the Great Plains, Australia, and Africa along parallel though slightly different lines, as the essays that follow in this book evidence. The same process impelled by a different dominant ideology has also been the center of Soviet policy toward Central Asian pastoral peoples, as Khazanov's essay demonstrates. In many places, an early step has been the removal of much original land and resource material from the reserve through damming or mining, both sponsored by the nation-state, its "Interior Department," and its client companies; the forcing of changes in education, religion, and family structures has come next. Yet later, some efforts to back away from the forcing of institutional change may have come in more enlightened countries as the forced changes have failed and efforts have been made, generally unsuccessfully, to accommodate both the tribal and the industrial in new institutions. This process, as it occurred in the Great Plains, the Saami Arctic, Australia, and Kenya, is described in the Hamilton, Morris, and Bekure and Pasha essays that follow.

The neocolonialist developmental interventions and their corollary institutions described in this section have had the effect of alienating group land and property using methods different from the "allotment" system described in the previous section. For example, the dam or mine controlled by outside corporations is often proposed to the indigenous group as part of an economic development strategy to benefit the group, but it may spoil vast areas of traditional lands. The indigenous corporation, in which the indigenous individual is a "stockholder," described by Anders in part 2 and by Bekure and Pasha below, may seem initially to serve corporate "tribal" interests yet become a vehicle for individualization and alienation. Even "outstations" designed to protect the traditional structure of Aboriginal life may become means of creating welfare dependencies and discouraging the pursuit of a traditional living on the land. This is not to say that such institutions are always "evil" and created by the dominant society with malice aforethought. They may be thought of by both indigenes and colonialists as transitional strategies, designed to act as buffers in the zone between indigenous and European based. However, they generally reflect the overwhelming power of the market mechanisms that move semiarid regions from a "subsistence" to an export or "cash crop" system as described by Bennett in his essay. In the extension of international market mechanisms to semiarid zones, a central problem in efforts to create new institutions that mediate between colonist and indigene has been that the buffering institutions have almost always been created by the colonists and subtly or blatantly serve their interests.

As Morris points out in his essay, no international protection—that is, nothing recognized as binding by the nation-state—exists to allow indigenous people to negotiate from a position of parity. Little in the World

Court, the United Nations, or the various international development agencies really protects such peoples at present, though the international agencies are making gestures in the direction of reform. Since these international agencies were created by the nation-states, they have had to move slowly, and perhaps only the common interest of humanity in protecting itself from general environmental disasters gives them the power to move at all. The ubiquitous "Interior Department Bureau of Indigenous Affairs" controls indigenous policy and takes its clue from the nation-state and the international market. As all the essayists who write in part 3 make clear, we do not yet have a group of professionals who understand both indigenous law and culture and its European-based equivalent (though tribal colleges not mentioned in the essays in this section may be beginning to do this). The destruction of traditional institutions for using the land and governing its use represents an issue of the abuse of justice and right, but it is also an issue involving the preservation-for-emulation of the sustainable ways of life regarded as essential in *Food 2000*.

Chapter Seven

Hydroelectric Development
and the Human Rights of
Indigenous People

C. Patrick Morris

THE PROBLEM: NATIONAL ENERGY DEMANDS AND
TRIBAL ENERGY RESOURCES

Since the Age of Discovery and the European conquest and colonization of the tribal world, the resulting nation-states have asserted exclusive sovereignty over the aboriginal inhabitants of the lands they "discovered." As the nation-state has replaced the tribe as the dominant political reality in most areas of the world, the more remote frontier regions beyond the crush of colonial exploitation and national incorporation have become the remaining homeland for the world's estimated 220 million indigenous people (Hitchcock 1985; Heinz 1985, 90–113).[1]

However, with the growing internationalization of the world's economy and associated demands for new energy resources, these isolated "ethnic" groups have found themselves, their lands, and their natural resources forcibly "incorporated" into the economies of the nation-states. To defend these last remaining tribal areas, native "activists" and organizations have taken to the streets, national courts, and international tribunals to contest national laws that permit state confiscation and destruction of what remains of the tribal world (Heinz 1985; Bennett 1978).[2]

One of the more prominent issues to evolve from the resource struggle between indigenous people and the nation-state is hydroelectric development. Since the mid-1960s, an unprecedented increase in hydroelectric development has occurred throughout the world. Many experts predict that by the year 2000, all or nearly all of the world's great rivers will be dammed and 66 percent of the world's stream flow impounded (Petts 1984; Mermel 1983).[3]

As a result of this global rush for hydroenergy, hundreds of dams have been constructed in the Arctic, tropical forests, and semiarid and arid "marginal" lands still inhabited and claimed by indigenous people (Dunbar 1983; Morris 1985). These dams have inundated millions of acres of tribal lands and natural resources and precipitated the forced relocation of hundreds of thousands, and in the near future millions, of native people (Dam destruction 1984; Aspelin and dos Santos 1981).[4] As a result, various legal and political efforts by native people to control hydroelectric schemes have become an important test of the presumed capacity of international human rights standards and laws to protect indigenous tribal minorities.

This chapter presents information about the development of hydroelectric projects in "tribal areas" of the United States, Norway, and Mexico. In most cases these tribal areas are in arid or semiarid regions where the struggle between tribal groups and nation-states over water and water development continues to be a major political and economic issue. The cases presented here were selected to demonstrate that the economies of indigenous populations in "water poor" regions are especially vulnerable to hydroelectric development. In these arid and semiarid regions, dam building means the loss not only of aboriginal lands but of scarce water resources as well (Morris 1985). Our survey begins in the semiarid North American Great Plains.

THE UNITED STATES INDIANS' CASE AGAINST HYDROPOWER: THE AMERICAN GREAT PLAINS

Since the end of World War II, the American Great Plains have undergone extensive economic development as a result of federal water projects.[5] One of the most expensive and ambitious of these was the Missouri River Pick-Sloan Plan. Authorized in 1944, the plan called for constructing 107 earthen dams in the Missouri River states to irrigate 5 million acres of land and generate more than 1.6 million kilowatts of electricity (Lawson 1982, 20–21). The economic benefits of the plan to the northern plains were obvious—cheap hydroelectric power, water for irrigation, regional flood control, and recreation.

The key structures of the Pick-Sloan Plan were five gigantic earthen dams. One of the first constructed was the Garrison Dam, built in North Dakota on the Fort Berthold Indian reservation.[6] When the work was completed in 1962, the impact of the Garrison Dam made it painfully clear who would pay for the benefits of the Pick-Sloan Plan—the Missouri River Indian tribes.

The Garrison Dam flooded 152,360 acres—25 percent of the entire Fort Berthold reservation, including 94 percent of the tribes' ancient agricultural lands. In addition, 80 percent of the reservation population was forced to relocate (Lawson 1982, 50). As the Missouri River waters rose, old and young people alike cried as most of the early history and culture of the Three Affiliated Tribes went quietly under water (MacGregor 1949). But this was only the beginning.

Of the five major Pick-Sloan Project dams built, the Oahe Dam constructed near Pierre, South Dakota, exceeded even the Garrison Dam in its destructive impact. In fact, the Oahe Dam destroyed more Indian land than any other public works project in America (Lawson 1982, 50, 51–53). More than 160,889 acres of land were lost on the Standing Rock and Cheyenne River Sioux reservations, including nearly 68 percent of the grazing land. Approximately 37 percent of the Indian families on the two reservations were forced to relocate (Lawson 1982, 52–54). Soon after, the Fort Randall Dam, built near the Yankton and Lower Brule Sioux, flooded an additional 22,000 acres of tribal farmland, followed by the Big Bend Dam near Fort Thompson on the Crow Creek and Lower Brule reservations, which flooded another 21,026 acres of tribal land (Lawson 1982, 47, 52, 54). In all, the five Pick-Sloan dams inundated most of the Missouri River tribes' most productive farm and timber lands. However, most devastating of all was the loss of the "wild" lands with their much cherished traditional plant and game foods (Lawson 1982, 56; Shanks 1974). Three decades later, there can be little doubt that the Pick-Sloan plan is a major contributor to the 70–90 percent unemployment rates on the region's Indian reservations today.

While the Missouri River Sioux struggled to preserve something of their lands and lives, farther upriver in the state of Montana other Indian tribes were fighting their own battles against federal hydroelectric projects.[7]

On the Crow Indian reservation in southeastern Montana, another Pick-Sloan project was being promoted—the Yellowtail Dam. The only obstacle was tribal approval. To build the dam the Bureau of Reclamation needed approximately 6,000 to 7,000 acres of Crow land on the Big Horn River, but the Crow tribe refused to sell. Furthermore, the United States Bureau of Reclamation discovered that, as an executive branch agency, any attempt it made to condemn tribal lands would violate the government's role as federal "trustee" of tribal resources (Berkman and Viscusi 1973, 163). Undaunted, the Bureau of Reclamation decided to overcome this legal obstacle by political means and began to organize a coalition of Montana businessmen, a local judge, the state's United States senator, and officials from the Bureau of Indian Affairs (BIA) to pressure the Crow tribe to sell out or face condemnation proceedings (Berkman and Viscusi 1973, 166). To add a financial cost to their threats, the Bureau of Reclamation informed the Crows that if they contested the bureau's original offer of $1.5 million, the bureau would see to it that the tribe received only $35,000 for the land (Berkman and Viscusi 1973, 172). However, the Crow tribal chairman, Robert Yellowtail, hired his own land assessor, who set the fair market value of the land at $5 million, more than three times the offer made by the federal "trustee" (Berkman and Viscusi 1973, 167–68). But rather than pay the market value for the land and end the struggle, the Bureau of Reclamation pushed ahead with its plan for a less expensive solution.

For the next ten years the Crows faced various federal, state, and private groups, each threatening the tribe with condemnation. These efforts to force an agreement even included interference with tribal elections (Berkman and Viscusi 1973). Finally, Congress agreed to pay the $5 million sought by the tribe. However, before payment could be made, the Bureau of Reclamation again challenged the amount and initiated a court battle that delayed final payment for another six years. By the time the Crows received payment for the Yellowtail Dam site, they had fought a bitter and divisive nineteen-year "war"—first to preserve the Big Horn River and then to receive "just compensation" for lands they did not want to sell (Berkman and Viscusi 1973, 178).

This is not the end of the Yellowtail Dam story. After the dam was completed, the Bureau of Reclamation began selling water rights to the Big Horn River—without legal authority. These sales involved several of the nation's largest energy corporations: Humble Oil and Refining Company, Shell Oil Company, Kerr-McGee Corporation, Sun Oil Company, and Peabody Coal Company (Berkman and Viscusi 1973, 178). Ostensibly these water sales were promoted by the BIA as part of an agreement to encourage the sale of Crow coal to these energy giants. But, when the Crows requested water for themselves, they were told that only 110,000 acre-feet remained unencumbered (Berkman and Viscusi 1973, 179). The Crow, in effect, were offered what was left. Yet within a year the commissioner of the Bureau of Reclamation and the Billings regional director of the BIA stated that about 750,000 acre-feet of water from the Yellowtail reservoir would be available for industrial consumption (Getches, Rosenfelt, and Wilkinson 1979, 220). Water was available to private and corporate interests, but not to the Crow tribe.

With this final act of deception, the federal "trustee" had successfully prevented the tribe from obtaining any economic benefit from the Yellowtail Dam hydroelectric project. For the Crows, the Yellowtail Dam is a staggering loss—loss of land, loss of potential revenues from hydroelectric power, and finally, the permanent loss of irreplaceable water rights.

Hydroelectric development in other areas of the United States has followed a pattern similar to that in the Great Plains, with a similar economic and environmental impact on Indian tribes. In fact, in the arid and semiarid regions of the American Southwest and California, hydroelectric development has been an integral part of what might be called the theft of the century—the massive taking of Indian lands and water to support economic development for non-Indians (Bernardo and Whittlesey 1986; Robinson 1979).[8]

Between 1908 and 1927, 93 dams were built in the Southwest by the Bureau of Reclamation. By 1976, according to Robinson, the bureau had built "320 diversion dams, 14,400 miles of canals, 900 miles of pipelines, 205 miles of tunnels, 34 miles of laterals, 15,200 miles of project drains, 145 pumping plants, 50 powerplants, and 16,240 circuit miles of transmission lines" throughout the West (Robinson 1979, 108). This federal building boom included $1.3 billion in irrigation systems and $1.5 billion in hydroelectric power facilities (Robinson 1979, 108).

To accomplish these feats of engineering, the Bureau of Reclamation took more than 1.8 million acres of Indian reservation land, most for hydroelectric development (American Indian Policy Review Commission 1976, 19). However, after sixty-five years of federal and state water development in the American Southwest, in 1967 it was reported that less than 1 percent of the more than 21.5 million acres of Indian land in the state of Arizona was irrigated (American Indian Policy Review Commission 1976, 26; United States Department of the Interior 1982, 19).[9] Instead, the water had been taken by non-Indians through federal dam projects—and more projects are on the way (Hayes 1980, 22).[10]

The economic results of this massive violation of the Indians' legal rights can be seen everywhere throughout the American Southwest—Indian reservation underdevelopment and poverty resting conspicuously alongside the spectacular growth and wealth of the southwestern sun belt. Ignoring the obvious, economic havoc created by western water development, official and popular explanations for Indian poverty continue to rely on the presumed lack of economic ingenuity and work ethic among Indians rather than the lack of water and water development on Indian reservations.

The federally subsidized theft of Indian water and land in the American West has been so flagrant that a 1973 Presidential Water Commission Report was forced to admit that "billions of dollars have been invested . . . in water resources projects benefitting non-Indians, but using water to which the Indians have a priority right" (United States National Water Commission 1973). To realign a mismanaged remark by former secretary of the interior James Watt, the "failed socialism" of the Indian reservation has underwritten the "free enterprise" spirit of the American West (*Billings Gazette* 19 January 1983, 1).[11]

Outside the United States in the Scandinavian Arctic, a similar struggle has been under way between that region's indigenous people, the Saami, and national hydroelectric development.

SAAMI REINDEER VERSUS NORWEGIAN DAMS

Across the arctic expanse of Scandinavia and the northwestern corner of the Soviet Union live more than 50,000 Saami, the indigenous people of northern Europe.[12] These ancient arctic people are best known for their transhumance reindeer herding, although a majority of Saami have always been farmers and coastal fishermen (Brantenberg 1985, 35; Eidheim 1971). With the increased industrialization of Scandinavia since World War II, many of the remote rivers of the arctic forest and tundra have been used to produce hydroelectricity (Paine 1982, 1; Nordic Lapp [Saami] Council 1969, 128–29, 143–51). By the 1960s the Saami found their access to traditional hunting and reindeer pasturage hindered, then lost, as rivers were captured and entire valleys flooded to generate electrical power for growing industrial cities in the south (Morris 1980–82).

Norway, with its smaller population and substantial North Sea oil reserves, lagged behind Sweden and Finland in northern hydroelectric development. However, in 1970 Norwegian Hydro, a state-owned corporation, proposed a gigantic hydroelectric project for the Alta-Kautekaino River in northern Norway. After considerable public opposition, the original Alta-Kautekaino Project (AKP) was reduced in size and eventually approved by the Norwegian government in 1978 (Paine 1982, 3).

The Norwegian government's determination to complete the AKP, despite strong public opposition, seems to spring from two complementary national policies: economic decentralization and an effort to strike a favorable balance between domestic energy consumption and profits from oil exports. To minimize domestic energy consumption yet increase the net availability of domestic energy, the AKP seemed a logical and profitable choice. Furthermore, the AKP held out the promise that surplus electricity could be sold to Norway's oil-poor neighbors, particularly Sweden (Morris 1980–82). Other factors that led to approval of the AKP were historical and cultural.

Throughout Norwegian history, the southern region of the country has been the nation's urban and industrial center. In contrast, northern Norway has remained isolated and limited in human resources and industrial development. Norway, in fact, is gradually becoming two countries, one urban and industrial, the other comprising northern small farms and fishing communities.

To overcome the ecological and cultural inertia that seemed to be pulling Norway apart, the government launched a policy of economic "decentralization" to promote commercial links between the South and North (Morris 1980–82). The key to this ambitious reintegration of the nation would be northern development of cheap electrical energy to attract new industries and reverse the migration of northerners to the South. Unfortunately, the government's ambitious economic plans failed to consider what impact major hydroelectric development might have on the indigenous people of the North, the Saami.

The site selected for the AKP was the Alta-Kautekaino River, which flows through the Norwegian "county" of Finnmark, the aboriginal homeland of the Saami and the least-developed region in Norway (Brantenberg 1985, 27). This vast arctic region, dominated by flat, almost treeless tundra, is cut by several of northern Europe's most spectacular rivers. To an outsider Finnmark appears barren and inhospitable, but to the Saami and the reindeer the open tundra is the ecological link that has fashioned the destiny of both man and animal for millennia. In these northern "Saami areas," as the Norwegians call them, the Saami language is dominant. The few Norwegians present in the Saami areas are either state bureaucrats or seasonal vacationers who have purchased cabins for fishing or hiking bases (Paine 1982, 2, 62).

To understand the reaction of the Saami to the AKP, it is necessary to understand the ambiguous relationship that has always existed between the Saami and the Norwegian people and state. Although Norway continues to be recognized as a world leader in international human rights, the Norwegian government and people actively discriminate against the Saami. Under the guise of an assimilationist educational policy called "Norwegianization," the government has made every effort to expunge the Saami language and culture from the national culture. For example, until very recently the Saami language was not permitted to be used as an instructional language in the public schools, even when Saami was the only language the children spoke. Forced to learn in Norwegian, fewer Saami have succeeded in the public schools (Brantenberg 1985, 45, nn. 2, 5). Also, despite "official" denials, Saami still suffer discrimination in housing and public accommodations (Morris 1980–82; Eidheim 1971, 1985, 155–71). Yet the Norwegian government continues to reject the possibility that Saami "ethnicity" might underlie many of these evident problems.

With the Norwegian government's history of benign indifference to issues related to Saami culture, language, and land rights, one might anticipate that any major development in the North, like the AKP, would soon bring the Saami people into direct confrontation with the Norwegian state.

In the fall of 1979 groups of Saami began to gather outside the Norwegian Parliament in downtown Oslo. Many carried signs that stated "We Came First" or "We Won't Go," expressing their refusal to abandon their homes and reindeer grazing areas to the AKP. Soon a semipermanent camp of *lavvo* or Saami tepees was set up outside the Norwegian Parliament, and the government found itself under siege (Brantenberg 1985, 38).

Throughout the winter of 1980–81, Norway was a nation polarized by the AKP issue. In December 1980 a regional Norwegian court of appraisal, the Norwegian federal appeals court, rendered a four-to-three decision against a large group of protestors who had camped at Stilla, the site of initial AKP construction (Brantenberg 1985, 38). Yet the protests went on. Before the winter of 1980–81 ended, Norwegian gunboats were sent north to serve as "prison" ships for those arrested at Stilla. Eventually the police and military arrested several hundred protestors, and the local court imposed heavy fines (Magga 1985, 15–22; Eidheim 1969).

In Oslo, several Saami men announced a hunger strike against the AKP. Later several Saami women occupied the Norwegian prime minister's office, while others traveled to the Vatican and sought an audience with the Pope to dramatize the Saami's situation (Magga 1985, 11–15; Brantenberg 1985, 39). Before it was over, two Saami hunger strikers went into comas and a third suffered permanent physical damage (Morris 1980–82). These were extraordinary days for a nation that viewed itself as a world leader in human rights.

Made aware of their inability to deal effectively with the AKP situation, the government struck an agreement with the protestors. If the Saami hunger strike would end, the government would review those portions of the initial environmental study that failed to take into account the impact of the AKP on Saami reindeer (IWGIA 1981, 63–76; Magga 1985, 15–16).

On the legal front, the December 1980 appeal court's ruling against the protestors was appealed to the Norwegian Supreme Court, which gave its ruling on February 26, 1982 (Paine 1982, 51; Brantenberg 1985, 38).

The decision of the Norwegian Supreme Court is important because it raises a number of legal and political questions regarding the ability of national and international law to protect the interests of indigenous people, even when opposed by benign national governments like Norway's.

With great care the Supreme Court judiciously avoided the international legal implications of the case and ruled on narrower "national" legal grounds, that is, confirmation of the Parliament's power to regulate the Alta-Kautekaino River. In addition, the Court went on to say that "no Saami interests have been violated by the AKP," and, to avoid international law, Paine writes, "the Court did not consider the Saami living in the areas affected by the project to be an indigenous people under the terms of International Common Law" (Paine 1982, 98–99). The court battle had been lost, but the political one continues.

In less wealthy countries such as Mexico, the country of our next case, indigenous people are the "poorest of the poor" and must contend with intense economic pressure as a result of the overwhelming national demand for any form of economic development. In such circumstances, indigenous people are usually powerless to stop governments from the wholesale exploitation of their natural resources to promote national economic growth.

MEXICO. THE MAZATECS, THE CHINANTECS, AND THE MAYANS

Like many developing countries that seem chronically to linger between economic takeoff and permanent stagnation, Mexico has a mixed rural subsistence and urban market-oriented economy. To extend the industrial economy of the nation into rural areas and reverse the growing migration of peasants into overcrowded cities, Mexico initiated its own "TVA" type of rural electrification project in the 1940s (Barabas and Bartolome 1973).[13]

As part of this rural electrification effort, the Mexican government in 1947 created the Papaloapan River Commission (PPC) to coordinate and develop electrical power for over 46,517 square kilometers in the state of Veracruz and parts of the states of Oaxaca and Puebla (Barabas and Bartolome 1973, 4). The major hydroelectric facilities would be the Miguel Alema and Cerro de Oro dams. However, the area to be inundated by these dams was the ancient homeland of some 90,000 Mazatec Indians, 70 percent of whom spoke only Mazatec (Barabas and Bartolome 1973, 6). What happened to these indigenous landowners when the PPC project was developed can only be described as officially sanctioned corruption and violation of the Mexican national constitution.

According to the PPC's own impact statement, the Miguel Alema and Cerro de Oro dams would flood much of the Mazatec Indians' riverine farmland and homes, necessitating extensive relocation. The PPC solution was to designate "five zones, located 50 to 250 kilometers from the original Mazatec habitat" for resettlement (Barabas and Bartolome 1973, 6). However, only one of these zones would have access to lands to be irrigated by the dams. The others would remain virtually desert.

Once the relocation began, Barabas and Bartolome write, the Mazatecs discovered that the best lands had already been given "to employees of the Commission or to influential people who expected to benefit from the dam's irrigation district" (1973, 7). "As compensation, the resettled Mazatecs were offered roads, potable water, and electricity, but these promises were not fulfilled" (1973, 6). When some Mazatecs refused to move, the commission simply ordered the dam's floodgates to be opened. Others who refused to move found their homes set afire by commission employees (Barabas and Bartolome 1973, 6).

Not everyone suffered from the dams, however. The PPC project turned out to be highly profitable. In fact, according to Barabas and Bartolome, "between 1947 and 1969 more than 110 million dollars went into the area . . . [including] . . . the Ingenio San Cristobal, one of the world's largest sugar refiners," which along with several other sugar refiners now processes more than 98,000 acres of sugar cane grown on lands irrigated from the dam reservoirs (Barabas and Bartolome 1973, 6). Of those Indians who still live in the area, many now work as laborers for the sugar refineries.

The Mazatec situation is only one of many in Mexico and Latin America (Wilkerson 1984). On the Mexico Guatemala border, a major hydroelectric project is planned on the Río Usumacinta, the river that separates the two countries. It is estimated that this joint Guatemala/Mexico project will require the relocation of over 25,000 people, primarily Mayan Indians (Wilkerson 1984). Once in operation, the dam's reservoir will destroy the local milpas, erode the region's thin tropical soils, promote clear-cutting of the hardwood forest, and endanger a number of rare animal species. Finally, the dams will inundate the ruins at Yaxchilan, a pre-Columbian Mayan city (Wilkerson 1984).

CONCLUSIONS

Today many of the world's 220 million indigenous people live in "undeveloped" arid and semiarid regions within nation-states. In these "waterpoor" regions indigenous people have struck a delicate balance with nature—one that has until recently successfully avoided the degradation of the world's fragile arid and semiarid environments. However, during the past twenty years hydroelectric development has had a disproportionate impact on indigenous people and native environments around the world. As a result, the scarce water resources of indigenous people living in arid and semiarid regions are being impounded, diverted, then reallocated to national majorities (Goldsmith and Hildyard, n.d.).[14] What follows is inevitable: the local aboriginal economy is drowned by man-made lakes, and the newly created irrigated lands are transferred to national economic interests. Eventually local native plants and animals are replaced with those that have "commercial" value—berries with wheat, buffalo with cattle, and traditional local markets with national or international ones. Finally the once self-sufficient indigenous populations either are compelled to join the new market-oriented economy as landless laborers or are forced to migrate elsewhere.

The destructive impact of hydroelectric projects in tribal areas around the world suggests that the present system of national "trusteeships" over tribal people and their natural resources has failed to develop any meaningful form of self-determination for indigenous people. Instead, the national "trustee" system has perpetuated an onerous form of internal colonialism by the use of such agencies as the Bureau of Indian Affairs (BIA) in the U.S., the Department of Inuit and Native Development (DIAND) in Canada, the Instituto Nacional Indigenista (INI) in Mexico, the Department of Aboriginal Affairs (DAA) in Australia, the Fundaçao Nacional do Indio (FUNAI) in Brazil, and so on around the world. Seldom scrutinized publicly or required to conform to national or international laws, these fiduciaries of the tribal world have successfully "legitimized" the theft of millions of acres of tribal lands and natural resources, the forced relocation of entire cultures, the systematic destruction of native languages, religions, and social and political communities—even the unlawful imprisonment and murder of those who legally oppose national policies they believe violate the human rights of indigenous minorities.

The continuing failure of the system of national trustees suggests that it is time for the world to question the presumed capacity of the nation-state to safeguard the human rights of indigenous people—particularly if those rights conflict with national economic priorities (IWGIA 1987).[15] Some alternative form of protective governance is needed to support meaningful forms of self-determination for indigenous nations. To address this and other issues an extranational tribunal is needed to investigate national wrongdoing and to formulate safeguards that would prevent the use of national "native" policies to "legally" violate international law and the human rights of indigenous people. Currently the United Nations serves this function in disputes involving member nations. But this limited United Nations oversight has not yet been extended to "native" cultures, for obvious reasons—the nation-states would view such oversight as an infringement on their sovereignty. Nevertheless, within the past two decades the United Nations, at the urging of native people and various international human rights organizations, has extended NGO (nongovernmental organization) status to various representative indigenous peoples' organizations (Eide 1985). This is a beginning, but only that. However, without such international efforts it is unlikely that the tribal world will survive much beyond the present century.

NOTES

1. For our purposes "indigenous," "native," and "tribal" refer to societies that are aboriginal to the lands they occupy and whose government and culture have not as yet been completely incorporated into the nation-states within which they reside. Indigenous people are often, but not always, racially and ethnically distinct from the national majority.

2. Generally, what has emerged in international law is the principle that indigenous tribal groups have more than a moral claim to their aboriginal lands—they also have a legal one. Examples of international political organizations formed to promote the rights of indigenous people include the World Council of Indigenous People (WCIP), Consejo Indio de Sud America (CISA), and the American Indian Movement (AIM).

3. World dam construction peaked in 1968, then declined. However, with the OPEC oil embargo in 1973 this decline was reversed. Between 1965 and 1976 worldwide hydroelectric dam construction jumped 90 percent, from 350 dams per year to nearly 700.

4. The worldwide invasion of tribal lands by hydroelectric projects involves every continent and more than fifty countries. In this chapter I have focused on only a few cases involving arid and semi-arid regions. This does not mean I am unaware of what has happened at James Bay, Island Falls, and other areas of Canada, the disastrous Allegheny Reservoir Project and resulting Kinzua Dam in upper New York State, or the unprecedented hydroelectric development now under way in the Amazon (See Aspelin, Paul L. and Silvio Cullho dos Santos 1981). Beyond the Americas, there is the Narmada Valley Development Project in the Indian states of Gujarat, Madhya Pradesh, and Maharashtra, consisting of some 30 large dams, 135 medium dams, and more than 3,000 minor dams. It is estimated that this project will eventually destroy

nearly a million acres of land and make homeless more than 100,000 people, many of them "Adivasi" or "tribals." By the end of this decade the total number of people in India adversely affected by hydroelectric development may reach 2 million.

5. Michael L. Lawson's book (1982) is one of the best contemporary Indian histories written and deserves wide readership.

6. The Fort Berthold Indian reservation is the homeland of the Three Affiliated Tribes, the Mandans, Hidatsas, and Arikaras.

7. The Kerr hydroelectric dam on the Flathead Indian reservation in northwestern Montana was recently the center of a fight between the Confederated Salish and Kootenai Tribe of the Flathead Indian reservation, the BIA, and Montana Power Company over who would receive the FERC license to manage the dam.

8. As a result of the Reclamation Act of 1902, "federal water" for irrigation increased dramatically. By 1980 more than 11 million acres were receiving water from federal reclamation projects. It is estimated that more than $6 billion has been spent on water-related projects by the Bureau of Reclamation. Between 1908 and 1927, the bureau built 93 dams. For an example of federal duplicity in hydroelectric development see Mariella (1983), on the now infamous ten-year struggle by the Fort McDowell Yavapais to stop the Orme Dam. The resulting reservoir would have inundated approximately two-thirds of the 24,000-acre reservation.

9. It is important to note that Indian water rights in the Colorado River were determined by *Arizona v. California*, 1963, which decided that the Colorado River tribes were entitled to one acre-foot of water per acre of irrigable land. In 1979 the Supreme Court ordered a water master to quantify Indians' water rights from the Colorado. In 1983 the master awarded 1.2 million acre-feet to the tribes, but this was rejected by the Supreme Court because such an award "cannot help but exacerbate potential water shortage problems" for non-Indians (Hundley 1981, 34).

To obtain needed congressional support for water projects, some southwestern tribes have had to give up some of their water rights. The Navajo tribe, the largest in the United States, in 1957 agreed to waive its priority on the San Juan River to get the Navajo Indian Irrigation Project (NIIP) under way. This irrigation project was first initiated in the 1950s but to date little progress has been made (see Reno 1981).

10. In the American Southwest the Department of the Interior, Bureau of Land Management, Bureau of Reclamation, and Bureau of Indian Affairs have all been involved in the development and now leasing of the region's energy resources, including hydropower, coal, gas, and oil. As Hayes reported, "In 1964 the above government agencies and twenty-three investor and publicly owned utilities formed a consortium known as the Western Energy Supply and Transmission Association (WEST). The WEST consortium planned to construct and operate a coal-fired generating and coal gasification network to produce three times the electric power of the na-

tion's largest utility, the Tennessee Valley Authority" (Hayes 1980, 22). This power goes largely to Los Angeles, Las Vegas, Phoenix, Tucson, San Diego, and Albuquerque. Population projects for the Colorado River basin areas are estimated to number 90 million by 1990.

11. In a public interview broadcast January 19, 1983, Secretary of the Interior James Watt described Indian reservations as examples of the failure of socialism.

12. The more widely recognized term "Laplander" or "Lapp" is rejected by the Saami as a slanderous "ethnic epithet."

13. For the nearly 4 million indigenous Indian people of Mexico, the "unnationalized" rural subsistence economy continues to dominate.

14. See note 4. During the next decade it is estimated that more than 10 million people in India alone will be displaced by hydroelectric development, many of them Adivasi or tribals. Similar events are unfolding in the United States, Sweden, Finland, Canada, Brazil, Malaysia, Indonesia, Sri Lanka, New Zealand, and the Philippines.

15. In a potentially important move, the Brazilian National Constitutional Assembly may consider the question of indigenous rights in the new constitution.

REFERENCES

American Indian Policy Review Commission. 1976. Reservation and resource development and protection: Final report. Report of Task Force 7. Washington, D.C.: Government Printing Office.

Aspelin, Paul L., and Silvio Cuelho dos Santos. 1981. *Indian areas threatened by hydroelectric development projects in Brazil.* IWGIA Document 44. Copenhagen: International Work Group for Indigenous Affairs.

Barabas, Alicia, and Miguel Bartolome. 1973. Hydraulic development and ethnocide: The Mazatec and Chinantec people of Oaxaca, Mexico. IWGIA Document 15. Copenhagen: International Work Group for Indigenous Affairs.

Bennett, Gordon. 1978. Aboriginal rights in international law. *Occasional Papers of the Royal Anthropological Institute of Great Britain and Ireland* 37.

Berkman, Richard L., and W. Kip Viscusi. 1973. *Damming the West: Ralph Nader's study group report on the Bureau of Reclamation.* New York: Grossman.

Bernardo, Daniel J., and Norman K. Whittlesey. 1986. The historical setting for irrigation in the West. In *Energy and water management in western irrigated agriculture,* ed. Norman K. Whittlesey. Studies in Water Policy Management 7. Boulder, Colo.: Westview Press.

Brantenberg, Odd Terje. 1985. The Alta-Kautekaino conflict: Saami reindeer herding and ethnopolitics. In *Native power: The quest for autonomy and nationhood of indigenous people*, ed. Jens Brosted. New York: Columbia University Press.

Dunbar, Robert G. 1983. *Forging new rights in western waters*. Lincoln: University of Nebraska Press.

Dam destruction—the case against superdams. 1984. *The Ecologist: Journal of the Post Industrial Age* 14, nos. 5/6 (Cornwall, U.K.: Ecosystems Ltd).

Eide, Asbjorn. 1985. Indigenous populations and human rights: The United Nations efforts at mid-way. In *Native power: The quest for autonomy and nationhood of indigenous peoples*, ed. Jens Brosted. Oxford: Oxford University Press.

Eidheim, Harald. 1969. When ethnic identity is a social stigma. In *Ethnic groups and boundaries*, ed. F. Barth. Oslo: University of Oslo Press.

———. 1971. *Aspects of the Lappish minority situation*. Oslo: University of Oslo Press.

———. 1985. Saami ethnicity. In *Native power: The quest for autonomy and nationhood of indigenous peoples*, ed. Jens Brosted. New York: Columbia University Press.

Getches, David H., Daniel M. Rosenfelt, and Charles F. Wilkinson. 1979. *Federal Indian law: Cases and materials*. St. Paul, Minn.: West Publishing Co.

Goldsmith, Edward, and Nicholas Hildyard. n.d. *The social and environmental effects of large dams*. Wadebridge Ecological Centre, U.K.

Hayes, Lynton R. 1980. *Energy, economic growth, and regionalism in the West*. Albuquerque: University of New Mexico Press.

Heinz, Wolfgang. 1985. International human rights and indigenous populations. IWGIA *Newsletter* 42:90–113.

Hitchcock, Robert K. 1985. Human rights and indigenous people. Manuscript.

Hundley, Norris. 1981. The West against itself: The Colorado River—an institutional history. In *New courses for the Colorado River: Major issues of the next century*, ed. Gary D. Weatherford and F. Lee Brown. Albuquerque: University of New Mexico Press.

International Work Group for Indigenous Affairs (IWGIA). 1981. Norway: Saami rights and the Alta-Kautekaino case. IWGIA *Newsletter* 27:13–76.

———. 1987. Brazil: Constitutional assembly reviews indigenous rights. IWGIA *Newsletter* 51/52.

Lawson, Michael L. 1982. *Dammed Indians: The Pick-Sloan Plan and the Missouri River Sioux, 1944–80*. Norman: University of Oklahoma Press.

MacGregor, Gordon. 1949. Attitudes of the Fort Berthold Indians regarding removal from the Garrison reservoir site and future administration of their reservation. *North Dakota History* 16(1): 31–60.

Magga, Ole Henrik. 1985. Are we finally to get our rights? On the work to secure the Saami peoples' rights. In *The quest for autonomy and nationhood of indigenous peo-*

ple, ed. Jens Brostod. Oxford: Oxford University Press.

Mariella, Pat. 1983. The political economy of federal resettlement policies affecting Native American communities: The Fort McDowell Yavapai case. Ph.D. diss., Arizona State University, Department of Anthropology.

Mermel, T. W. 1983. Major dams of the world—1983. *Water Power and Dam Construction* 8:35.

Morris, C. Patrick. 1980–82. Contemporary Saami politics. Unpublished field notes in the author's possession.

———. 1985. As long as the water flows . . . United States Indian water rights: A growing national conflict. In *Native power: The quest for autonomy and nationhood of indigenous people*, ed. Jens Brosted. New York: Columbia University Press.

Nordic Lapp (Saami) Council. 1969. *The Lapps today in Finland, Norway and Sweden*. Oslo: University of Oslo Press.

Paine, Robert. 1982. *Dam a river dam a people*. Copenhagen: International Work Group on Indigenous Affairs.

Petts, G. E. 1984. *Impounded rivers: Perspectives for ecological management*. New York: John Wiley.

Reno, Philip. 1981. *Mother Earth, Father Sky, and economic development: Navajo resources and their use*. Albuquerque: University of New Mexico Press.

Robinson, Michael C. 1979. *Water for the West: The Bureau of Reclamation, 1902–1977*. Chicago: Public Works Historical Society.

Shanks, Bernard D. 1974. The American Indian and Missouri River water developments. *Water Resource Bulletin* 10:573–79.

United States Department of the Interior. 1982. *A year of progress: Preparing for the 21st century*. Washington, D.C.: Government Printing Office.

United States National Water Commission. 1973. Water policies for the future. In *Final report to the president and to the Congress of the United States*. Washington, D.C.: Government Printing Office.

Wilkerson, S. Jeffrey. 1984. Mexico and Guatemala: Archeological and ecological implications of proposed hydroelectric projects on the Río Usumacinta. IWGIA *Newsletter* 38:75–89.

Chapter Eight

Culture Conflict and
Resource Management in
Central Australia

Annette Hamilton

Central Australia is one of the world's most difficult living environments, with an average rainfall of 130–250 millimeters (map 8.1). Temperature variations of 30° to 120° occur annually, and long droughts are frequent. Irrigation is limited to bore water pumped from the artesian basin, which is insufficient for agriculture. Although classified as desert, the region does not resemble African and Asian desert but has a highly specialized flora and fauna able to survive both sudden rains and long droughts (map 8.2).

Aborigines settled in this area at least ten thousand years ago, whereas the first white explorers and settlers came only one hundred years ago, displacing the Aborigines from much of the most fertile territory. The significance of this settlement and displacement becomes clear if one examines the meaning found in the landscape by the Aborigines, the history of white intervention and its mythology, the effects on the Aborigines and the environment, and particularly, recent efforts to create means of accommodating the differences between the two cultural, social, and economic systems.

The Aboriginal inhabitants of Central Australia developed specialized skills for survival. Divided into a number of linguistic groups with a series of dialects, the Aboriginal population of the central area might once have numbered about 10,000 to 15,000 people.[1] Although marked differences existed, and still exist, among the various groups in social organization and ritual practices, they all lived by hunting and by gathering the products of the

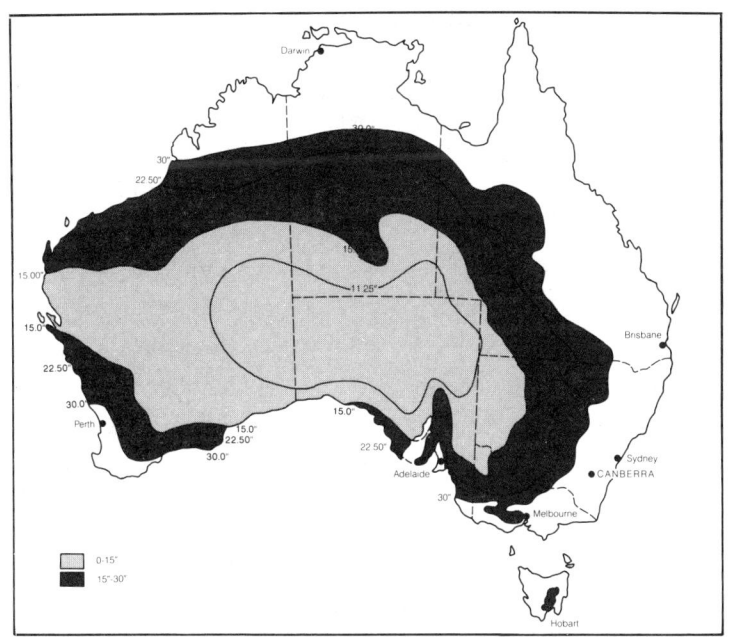

Map 8.1 Annual rainfall in Australia.

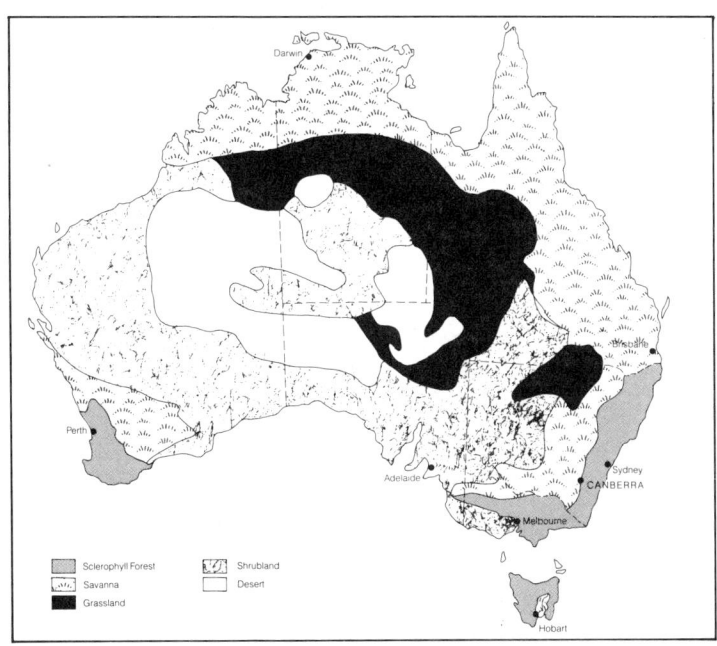

Map 8.2 Natural vegetation in Australia.

land. Men hunted large animals using spear and spear thrower; men, women, and children all gathered smaller protein and vegetable foods. Women gathered grass seeds in season and ground them into flour for a bread baked in the embers, where all cooking was done. Shelter consisted of base camps, including substantial shelters made of boughs and grass or temporary shelters of bark over saplings (Meggitt 1962; Gould 1969; Myers 1976, 1982; Kimber 1984). The technology of Aboriginal culture was simple in the extreme: wooden tools including spears and spear throwers, digging sticks, shovel scrapers, and large and small bowls—all made with stone and fire— stone tools, vegetable resin, and thongs made from animal sinews. Human hair spun into string was worn as decoration or used for tying up parcels; ochred and fatted, it constituted an important part of the ceremonial exchange cycle.

Though observers of Aboriginal society have characterized its mode of adaptation as "passive" or even "parasitic" (Elkin 1951), recent research has shown that the conventional depiction of Australian hunter-gatherers has missed their techniques of storage (Allen 1974; Testart, n.d.), their complicated methods of food preparation by neutralizing poisonous substances (Beaton 1977), much of their technological apparatus such as fish traps (Lourandos 1980; Coleman 1982), and even their permanent stone dwellings in some areas (Flood 1983; 205–6). Above all, it missed how they used fire to markedly alter the natural environment, increase useful food plants and animals, and bring about a "control" that one scholar likened to a technique of "farming" (Jones 1969; Hallam 1975; Latz and Griffen 1978; Horton 1982; Lewis 1982).

However, it may be that the most far-reaching and ecologically powerful technology was of an immaterial kind. Based on his extensive knowledge of the Aranda-speaking peoples, the late Ted Strehlow argued that Aboriginal religious beliefs and practices ensured the reproduction of species not merely in their own ideas but in reality, by providing sacred sites where no one would hunt or seek water or disturb the natural species, and where animals and plants could survive and reproduce safe from human threat even in the most terrible droughts (Strehlow 1965, 143). Aboriginal people find their way unerringly through trackless wastes not because they understand how to navigate by the stars, but because they have encoded in their memories the names of all the watering places from one end of their living range to the other in song verses that tell of mythical travels of creative ancestral powers. These verses name the places where water and sometimes other resources will be found and include information on their physical appearance and the directions from one to another (Strehlow 1970, 93).

Traditional practices regarding elements we would regard as "technological" or "adaptational" are also encoded within belief systems as part of the mythicoritual world of the Dreaming and its Law: animals must be cooked in their skins; certain parts must be distributed to certain kin of the hunter, who must himself take the worst part or none at all; men should set off on food-gathering or hunting expeditions in one direction, women in another. All these beliefs can be shown to have consequences in terms of ecology and adaptation.[2] Whether these practices should be taken to constitute a program of resource management is a matter of approach and definition. Those who argue that only consciously held understandings of the relation between means and ends should be taken as "management" will be opposed by those whose view of human cultural adaptation includes the possibility that a "good" set of lifeways will result in a balanced and viable relationship between human beings and their environment even if the humans are not actively aware of means-ends relations. The indisputable fact is that the Central Australian Aboriginal adaptation was one that could sustain a continuing and reasonably large human population in an environment so marginal as to barely support a few European families in the way Europeans, even with all their technological superiority, customarily expect to exist.[3]

For traditional Aboriginal people what the world *means* was set out for all time by the activities of creative ancestors who wandered across the countryside, creating species, naming places, engaging in interactions with one another, and leaving behind storehouses of creativity that human beings can tap into to ensure the reproduction of all species, including themselves. This creative period is commonly referred to as "the Dreaming," and though it is in one sense situated in an immensely ancient "time," it is also permanently present—immanent—to be tapped by humans through ritual and spiritual activities. Water holes, ridges, groups of trees, claypans, groupings of stones—all are visible marks of the creative period and testify to the truth of the Aboriginal version of events. The creative heroes slumber still beneath the ground; the giant water snake lives in deep pools and manifests itself as a rainbow; women become pregnant because they encounter spirit children waiting to be reborn.[4]

Human health depends on maintaining the cosmic order by guarding the sacred places, keeping the rituals, watching ceremonial performances, and performing male initiations.[5] If those who are responsible for the dreaming tracks are derelict, all other Aboriginal people with links to those same totemic heroes will also be weakened and will hold the custodians responsible. Animals and plants, formed from the Aboriginal people's creative power, share an identity with them; for example, kangaroos are born from the same creative power that animates the fetus of a human whose totemic affiliation is kangaroo. Even ice, mosquitoes, flies, and illnesses have their counterparts or source in the ancestral times and in their turn must be controlled by ritual attention involving song verses, body painting, and sometimes ground painting, dancing, and bodily sacrifice.

Hence, points in the landscape function as signifiers for Aboriginal people that provide meaning for their bodies, their lives, and their own creative powers because of the mythologies that knit "the real" together with "the imaginary" and assert a commonality between human and landscape. There is no call for modification or transformation of one in the interest of the other; they are in a relationship of mutual dependence, and though human beings must maintain the relationship and the natural creativity of "the real," they must also avoid any damage to its essential stability, a view apparently not undermined by the European presence in the last century.

The first white settlers of the Center, who came immediately after the explorations of the 1860s–70s, obviously did not adhere to this vision of mutual dependence but took the best pieces of the country—"picked the eyes out of it"—appropriating the best waters (see Rowley 1970 for a general history). Valleys with semipermanent water holes that once supported hundreds of people now became the home first for hundreds of sheep, then for cattle, together with one or two white men—many of whom took Aboriginal women as their consorts (Strehlow 1978). The Aboriginal people everywhere mounted a vigorous resistance to their dispossession—but since they were considered by British law to be Crown subjects, this action of self-defense was defined as criminal activity, not war, and hence they were shot or incarcerated; no treaty rights, rights of Aboriginal "nationhood," or prior rights of occupancy were recognized (see Reynolds 1988). In central Austra-

lia the first decade of white settlement was a pleasing one for the intruders—
it coincided with a series of good seasons, with good rainfall, and the native
grasses and plants provided excellent fodder. The crash came in the 1890s,
when drought followed drought and prices for cattle dropped in the Australian market. By then, too, large areas of land were denuded of the native
flora, much of which never recovered from decades of overgrazing in good
seasons (Hartwig 1965).

After the white takeover, many Aborigines living in remote desert regions remained autonomous until the 1930s, some even until 1984. But most
formed camps around the European stations, surviving on occasional labor,
handouts, and traditional economic pursuits. The period was characterized
by early death, disease, and malnutrition. Influenza and measles decimated
the surviving population. The rifle and the metal digging stick replaced traditional food-gathering equipment, steel tools replaced stone ones, and
most important, transport changed. Early white Center settlements were
serviced by Afghan camel trains used for long-distance transport, and the
Aborigines soon adopted camels. Two or three camels would be used to
transport children, old people, and new material goods on long expeditions
around traditional traveling routes. Although people inevitably returned to
base camps near European settlements, the camel permitted an extension of
nomadism hitherto restricted by availability of surface water, and it continued to do so until the motor vehicle supplanted the camel and, by the early
1970s, replaced it. In spite of the conditions facing them, Aboriginal people
survived and reproduced, and Aboriginal culture, though it underwent radical transformations, did not collapse, but adapted. By the Second World
War not only was the mixed-blood Aboriginal population increasing, but
the full-blood Aboriginal population began to regain its strength.

After the 1940s changes came apace, but they did not all go in the same
direction. In the 1950s most Aboriginal people in the Center and the North
became wards of the state, subject to complete surveillance on reserves and
settlements staffed by a motley crew of whites (Rowley, 1970). In the 1960s
the "Aboriginal problem" began to be recognized as a national issue, and
the ideal of assimilation was promulgated. Thousands of dollars were
poured into schools to make Aboriginal children become Anglo-Australian
adults within one generation, a policy whose futility became apparent
within the decade. Strong Aboriginal voices began to be heard in the South,

arguing for "Aboriginality" and treating assimilation as a form of genocide.

The 1972 Labour government gave Aboriginal affairs a new political importance by assuring Aborigines of their right to a separate non-Anglo existence and recognizing Aboriginal prior occupation of the country in the form of a federal Land Rights Act promulgated in the Northern Territory. Some, but not all, states enacted their own legislation for recognizing Aboriginal rights to land, but only the Northern Territory—a territory and not a state—is subject to the Commonwealth law and thus falls under the federal Land Rights Act. Since central Australia's Aborigines straddle three state boundaries, the Northern Territory, Western Australia, and South Australia, only two of which have recognized Aboriginal rights in land, not all their land is secure, since Western Australia consistently refuses to recognize Aboriginal land rights.

The Northern Territory legislation required Aborigines to lodge land claims to vacant Crown land and substantiate them through a rigorous court procedure demonstrating that as a group they constituted the traditional Aboriginal owners (Peterson 1981; Maddock 1981), and vacant Crown land has generally been so marginal that Europeans have not wanted it. Hence lengthy, expensive court cases have given the Aborigines inalienable freehold title to much economically marginal land. A clause in the federal Land Rights Act also permits Aboriginal people having European title to an area to convert it into Aboriginal land. Though productive and valuable areas have been obtained in this way, this avenue to landownership is opposed vigorously by the Northern Territory government on the grounds that "too much" of the Territory is going into Aboriginal hands and will become "unproductive."[6]

The many changes in power and position of Aborigines in the past two decades have brought dramatic alterations in the institutional relationships mediating between them and the non-Aborigines. The character of the European-based occupation of Central Australia is changing, and though much of Central Australian country is still devoted to cattle raising, old family-based enterprises are disappearing rapidly in favor of managed stations with absentee owners, often wealthy southern farmers or companies or foreign owners—for example, the sultan of Brunei. Few Aborigines now live on cattle stations. After the Land Rights Act, former reserves were handed over to Aboriginal ownership; under the "self-determination" policy infrastructural support for "outstations" became available so that many Aboriginal people moved away from towns and settlements and established extended family-based locations with minimal services but maximum social autonomy, often at places close to traditional religious sites and isolated from non-Aboriginal social pressure. Little or no economic activity is possible on these outstations, and people live largely on conventional welfare subsidies—pensions for old people and widows, unemployment benefits, family allowances, and so on. The outstations contain the most "traditional" elements of Aboriginal society (Coombs 1974; Wallace 1977; Meehan and Jones 1980; Nathan and Japanangka 1983).

Aborigines who have not opted for this Aboriginally "conservative" choice may live and work in Alice Springs, the single large town and the service center for the vast central Australian hinterland. With a population of 20,000, of which about 3,000 people are Aboriginal or part-Aboriginal, Alice Springs is the magnet for the Aborigines of the Center, providing jobs and housing for some of the better educated and a place to drink and misbehave for many from remote settlements and outstations. Traditional methods of social control are absent or ineffective, and the white man's law prevails in the form of police who arrest Aborigines with alacrity for committing offenses such as drinking alcohol within two kilometers of licensed premises (hotels). For some, especially people whose families have lived in and around Alice Springs for a generation or more, it is their only home. Though cut off from traditional tribal homelands, and from much traditional ritual and religious life, these Alice Springs Aborigines are vocal and active and understand how to get the most out of the European presence in the area while maintaining an Aboriginal identity. They are active in fostering initiatives for the betterment of Aboriginal conditions, and many support efforts to gain additional land rights both in town areas and in more remote regions.

In spite of European technology, nobody survives fully on central Australian land. Aborigines still gain much from it, especially at the outstations, but remain dependent on income from the Australian welfare system. White entrepreneurs in Alice Springs make a good living but depend on the public servants, teachers, and other townspersons whose income is paid by the Australian taxpayers. The cattle station managers and their families gain a livelihood and, in good years, considerable wealth from cattle raising, but this is negated in bad years, and many cattle enterprises are in fact tax shelters.

Although European agropastoral adaptations have not been a great economic success, two additional developments—mining and tourism—have had a considerable impact in recent times. Small-scale mining of gold, tin, and wolfram has existed for many years. In the past decade vast reserves of gas and petroleum have been located and tapped, and gold and uranium deposits appear to be developable with new technologies, promising considerable returns. Now both Australian and overseas-based consortia are doing the developing. Tourism has also expanded and promises good returns for hotels and small businesses, with facilities mushrooming in Alice Springs and in scenic areas outside it. These developments have affected the Aborigines in some obvious and some subtle ways (Hamilton 1984).

The net effect of this century of interaction has been the production of a complicated mosaic of Aboriginal life and of marked disparities between Aboriginal people. Whereas Aboriginal social distinctions once recognized language, marriage, and ritual practices and occupation of particular tracts of land as crucial, today's social distinctions recognize residence, education, knowledge of English, employment, and access to grants of land under the Land Rights Act. The last distinction is further enlarged in consequence of the royalty provisions in the Northern Territory Land Rights Act, whereby those Aboriginal people who are accepted as Traditional Owners or members of affected communities are entitled to a share of royalties from mining enterprises on their lands (see Altman 1983 and map 8.3; cf. Berndt 1982; Von Sturmer 1982; Howitt and Douglas 1983).

The objective effects of the past fifty years of human occupation of the central Australian landscape are extremely complicated and not yet fully apparent. The overgrazing of the land, the erosion of the soil, mining scars, and the disappearance of animal species are the most obvious effects (Rolls 1969; Hamilton 1984), but the contemporary Aboriginal population has also had a powerful impact on the local landscape by creating tire-mark roads, denuding woodland areas, and destroying animal life.[7] Aborigines continue to hunt animals, but often their rifles are in poor condition so that wounded animals may escape and perish without providing food for anyone but wild dogs and scavenging birds. Other animals are then hunted until one is killed outright.

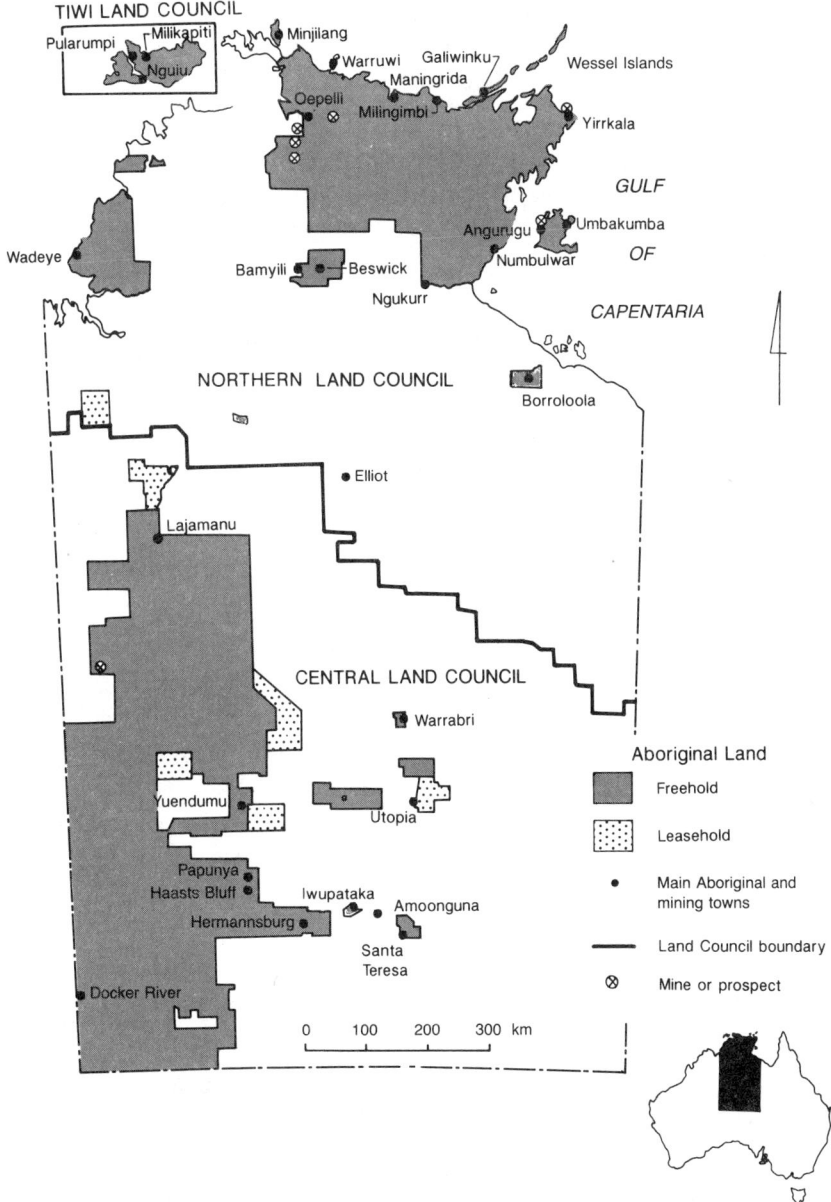

Map 8.3 Mining of Aboriginal land in the Australian Northern Territory in 1983.

Future change will be defined by technology but also by three sets of myth: the Aboriginal, the northern, and the southern Australian.[8] I defined the Aboriginal view in a simplified way early in this chapter as based on a mutual dependence between natural and human forces.[9]

In contrast, the non-Aboriginal developers of the Northern Territory have little understanding of the Aborigines[10] and tend to take a sort of "John Wayne" view that the frontier is a hostile environment, requiring radical technological control, and that it is also Australia's frontier against Asian hordes (Franklin 1976; Evans 1975, 123; Curthoys and Markus 1978). The same people tend to feel that the South is effete and even socialist, an impediment to further northern development—despite the fact that most northern development schemes have been funded from the South. Growing cotton, rice, and soybeans has failed (Mackenzie 1980), and only cattle, mining, and tourism have succeeded or promise success.[11] Aboriginal land rights, the Aborigines' reception of standard welfare benefits, their opposition to uncontrolled development and to further cattle raising are perceived as an offense against the developmental ideology of the northerners.

In contrast, southerners often perceive the North as crude, anticonservationist, and excessively "macho." The Northern Territory government's insistence on mining in national parks has aroused outraged response in some quarters, whereas the mining interests tend to view the problems as soluble with strict environmental protection plans (Saddler 1980).[12]

How do Aborigines fit in with southern ideas of "conservation" in those places where they are the only ones fully "at home"? Uncomfortably. Although conservationists embraced Aboriginal causes a decade ago—on the grounds that Aborigines were the natural conservators of the environment and knew best how to manage it—the conservationist credentials of Aborigines are questioned by southerners when they see Aboriginal camps littered with broken-down cars and car parts, with ruined buildings, no trees or shade, and crude shelters made of corrugated iron and canvas. Aborigines want to hunt the very animals that conservationists want preserved, and they use rifles and vehicles that destroy the natural environment as surely as any tourists. Thus the coalition between Aborigines and conservationists has largely collapsed.[13]

The three views caricatured here have their political expressions at local, state, and national levels. All rest on a set of apprehensions of reality that are seldom consciously articulated into principles of value, each party holding to the evident truth of its own deductions. The most obvious conflicts occur when there is need to negotiate over some concrete development where the "value" hierarchies come into play and institutions for resolving conflict are unsatisfactory.

One example of this relates to the question of compensation for disturbances on Aboriginal land: for example, a new mining venture. The local white community is always fully behind such ventures, as heralding further "development." The southerners become involved through the corporate structures of the developers, boards of management, lawyers, and negotiators. Southern conservationists, and often antiuranium groups, oppose them. The Aborigines are required to consent to such developments and agree on terms. How are they to calculate the "value" to them of such disturbances?

From the Aboriginal point of view, this value depends on the nature of the development, where it is, and what level of local transformations may be expected. Is the proposed development going to affect a large and significant sacred site or complex? Will it involve going under the ground in a limited way (oil wells) or in a large way (open-cut mining)? Will it involve roads that cut across traditional ritual pathways for carrying ceremonial objects from place to place? Is it in a place where known dangers lie (an illness dreaming place)? Will it involve towers that may overlook vast areas including sacred sites and tracks? Will it cut across dreaming tracks in such a way as to "break" them? Which Aboriginal custodians should properly be asked to make the decision? All these questions involve complicated matters of internal Aboriginal ritual and social politics, often involving people who live hundreds of miles away from each other.

In contrast, European decision-making requires "nominees" to make decisions on behalf of others. Though Aborigines have lawyers and other white advisers to assist them, the views of these experts may also diverge: the lawyer's main interest may be to secure the optimum financial return; the white adviser may comprehend the implications of such development for the future populations in a way the Aborigines may not and may wish to discourage or encourage it accordingly; some Aborigines may see the venture as an opportunity for financial independence while others may oppose it because they value their country and its sites more highly. And these differences may lead to disputes, accusations of sacrilege, or even death.

Should a venture proceed further, other problems may arise—world markets, financial terms, and changes in legislation may affect overnight the "value" of any newly developed enterprises. Explaining all of these matters to people whose first language is not English is a daunting task, and the overarching view that money should compensate for anything is not always shared. As J. Biendurry, of the Kimberley Land Council in Western Australia, said in 1980: "We want the land, but not to make money. That's very important. We're not going to muck up our countryside . . . just for making money. We want our countryside just like it was a few thousand years back—natural countryside" (Australian Land Rights Support Group Newsletter 7, 1980).

Whereas Aboriginal values stress continuity and the perpetuation of the Law, white values stress dynamism and the ability to transform the Law, and Aborigines know that the decision of one day may not be the decision of another—that governments change, as does government law. A law that gave them rights over their land in the 1970s may take them away because those rights do not sit comfortably in the 1990s.

The 1970s government self-determination policy assumed that Aborigines lived in communities where consensus decision making could take place. Many Aborigines also hold to this ideal, but the reality is often different (see Von Sturmer 1982). Cleavages within the Aboriginal population of central Australia may not permit "traditional" decision making; lack of qualified Aboriginal people who can negotiate as lawyers and advisers complicates matters, and "qualified" Aborigines are immediately recruited into the bureaucracy designed to deal with "Aboriginal affairs" and become immobilized. Thus fully informed decisions are seldom possible.

Allowing uncontrolled European access to Aboriginal (and all other) land is equally not a solution. The Aboriginal population is growing, and outside the urban areas Aborigines will remain the dominant group. Whatever happens, the assumption of rights to land and its resources by Europeans will no longer be tolerated without opposition. And there is the question, Development for what? Who will profit from it? How much can the southern taxpayers be asked to bear in the name of northern development? Attempts are being made to develop joint management strategies, for example, over national parks (Ayers Rock is a prime example). Other Aboriginal industries are supported by the central government with differing levels of success. For example, the cattle industry's problems are even worse at Aboriginal cattle stations, which lack the capital infrastructure and possibly the economic motivation to maintain fences and bores, keep proper veterinary surveillance, and market effectively. Sometimes "good ideas" are given a trial—Aborigines of northwestern South Australia are planning to export wild camels to Morocco. However, all such efforts to meld the European economic system with Aboriginal culture, codes, and values run against problems that lead back to incompatibilities in the systems: for example,

1. The lack of an adequate capital base for Aborigine-controlled developments and the various governments' refusal to come to terms with the question of who should provide it.

2. Lack of Aboriginal expertise in control and management of capital as well as the undertaking of new economic initiatives.

3. Lack of vision and coordination from any source regarding possible viable alternatives acceptable to all parties, allowing for the divergent codes and practices that follow from them.

4. Limited or no evaluations by non-Aborigines of what an "economic success" would be, or even what constitutes a valuable and viable lifestyle for Aboriginal people.

5. The prevalence of a "romantic" vision among non-Aborigines (and some Aborigines too) of what Aboriginal culture is really about, and a consequent refusal to see the possibility of radical transformations as anything other than destructive.

Finally, the main difficulty seems to be a complete lack of cooperative objective analysis between Anglo-Australian professionals and Aboriginal people as to what the possibilities are for preserving a satisfactory human life in the arid and semiarid zones. Embedded as all parties seem to be in a network of meanings that scarcely overlap, sustained by complex and mutually incomprehensible mythologies, resentment and anger grow with each passing year. Finally, it comes down to power. The hegemonic domination of Anglo-Australian society is increasing through its control of legislation, land, and resources; for Aborigines to respond, to maintain their codes and meanings, becomes increasingly difficult, especially as national support dwindles and life itself becomes precarious for many because of illness, alcoholism, poverty, despair, and frustration.

A genuine northern development would involve a redefinition of mythoi such that all parties could recognize the legitimate interests of others. It is my experience in many instances of negotiation that Aboriginal people are both willing and able to do this, and indeed can suggest quite remarkable compromises that meet the needs of all parties. However, these solutions still do not fit the inherent although "unconscious" symbolic parameters of a "proper development" in European terms, and they are often regarded as quaint impracticalities that demonstrate how "unrealistic" Aboriginal people are. The term "alternative" has taken on extremely negative connotations in contemporary Australia. To suggest that a different approach to "development" might be both more practical and more suited to a long-term human adjustment is greeted in many quarters as unpatriotic. In any case the short-term view is always considered the priority because of "rational" economic values. This must be an extraordinary perspective to Aborigines accustomed to seeing the continuation of the past into the present as the essence of orderly human life. Only a conscious act of will and intelligence from the non-Aboriginal people, and a profound reconstruction of contemporary Aboriginal existence, could offer promise of a harmonious development of the arid zones. Unfortunately the history of expanding settler colonies does not offer much cause for optimism.

NOTES

1. The boundaries of "central Australia" are problematic; the physical features and Aboriginal culture zones extend almost to the coast of Western Australia. Further, there are great difficulties in accurately reconstructing the size of the original Aboriginal population anywhere in Australia (see Smith 1980).

2. No published work comprehensively analyzes this "immaterial" aspect of Aboriginal technoenvironmental adaptation (but see Peterson 1972). However, there is considerable literature concerning Aboriginal population control mechanisms, spacing, territoriality, and local organization (see, for example, Peterson 1975, among many others).

3. The antiquity of Aboriginal occupation in the central desert is not firmly established owing to the paucity of archaeological investigations. A date of about 10,000 B.P. is generally accepted, based on the excavations by Gould at Puntutjarpa rockshelter (Gould 1968). Much earlier dates of occupation have been established in more

favorable parts of the continent; for example, the Devil's Lair site dates firmly as earlier than 30,000 B.P. (Dortch 1979; for further information see White and O'Connell 1979).

4. Numerous anthropological reports detail the variants of Aboriginal beliefs concerning "the Dreaming," the creative period, totemic beliefs, and so on (see, for example, Strehlow 1947; Stanner 1964).

5. For a discussion of the relation between women's ritual performance and the maintenance of human health, see Bell (1983, 145–62).

6. This occurs because Aborigines, once in control of their land, do not generally maintain it with an eye to economic maximization, but rather allow it to revert to "natural vegetation."

7. Populations of euro (*Macropus robustus*) and red kangaroo (*M. rufus*) have benefited from changes in pasture brought about by grazing. However, this may be a short-term consequence. In the long run, the introduction of ruminant herbivores may result in the extinction of marsupial fauna (see Newsome 1980).

8. The following discussion of mythoi is of necessity brief, schematic, and oversimplified. A full analysis would require an extensive discussion in terms of symbolic systems and their interplay with power relations. Even the depiction of "three parties" overlooks the complex interplay of codes and meanings within factions of the various parties. This limited discussion is put forward as an indicator of the way a symbolic analysis of the production and control of meaning is a necessary element of all discussion of such "conflict" situations. In many cases analysis gives priority to political and economic relations and relegates systems of signification to a theoretical periphery.

9. For an Aboriginal perspective on the relation between people, land, and resources, see Lhanhupuy (1982). He points out that "non-Aborigines require a total anthropological and sociological comprehension of Aboriginal life in order to understand what responsibility to the land means."

10. There are of course many non-Aborigines (apart from scholars) who do have a genuine understanding of, and sympathy for, Aboriginal religious beliefs. Some are ministers of religion, schoolteachers, and long-term residents with close associations with Aboriginal people. However, many who have a sympathetic attitude find themselves constrained by the social environment in which they live and work, and they soon cease to express their understanding to other non-Aborigines, thus confirming the hegemonic codes of remote Australia.

11. "Any agricultural product which can be produced in the region can be produced in Queensland or northern New South Wales at lower cost as wages and other costs are much lower in the latter region" (Davidson 1980, 82).

12. The mining industry, through the Australia Mining Industry Council, has

mounted a concerted campaign against both the Northern Territory Land Rights Act and the proposed national land rights legislation, on the grounds that Aboriginal rights to veto development hinder exploration and mining. This has fed into the national political arena in some remarkable ways, particularly in relation to internal Labour party politics. Hugh Morgan, president of the Mining Industry of Australia and a spokesman for the "New Right," has expressed his industry's view that the efficiency and productivity of the mining industry constitute an absolute benefit to the Australian community in terms of the industry's contribution through taxation, provision of infrastructures and so on, and that Aboriginal rights hinder this to the detriment of all Australians. Underlying this position is the argument that Australia is "one nation"; this view rests on the belief that the continent was legally occupied, which is precisely what Aborigines argue against. Thus descriptive terms ("one nation") come to carry a heavy burden of meaning, linking into a chain of significations connecting legal occupation of the land by non-Aborigines, the dominance of "assimilation" as a correct model for Australian development, the illegitimacy of "minorities," the priority of Anglo-Australian values, and the legitimacy of business and capitalistic development. Thus the term connotes a kind of spiritual sanctity that only the deviant would question. On the other side, the term "sacred site" has come to carry a similar chain of significations.

13. At a deeper level, the visible presence of Aborigines seems to threaten important cultural values of the Australian way of life itself. Aboriginal perceptions of "propriety" include behavior, in itself quite harmless, that nonetheless acts as a "challenge" or a "shock" to non-Aboriginal Australians. Use of public space provides a good example. Aborigines will sit on the ground talking, playing cards, feeding babies, eating, and drinking, in full view of passers-by. This is perceived as somehow "indecent" by members of a cultural system that regards public space as space to be traversed, not occupied. Many other examples could be adduced.

REFERENCES

Allen, Harry. 1974. The Bagundji of the Darling basin: Cereal gatherers in an uncertain environment. *World Archaeology* 5:309–22.

Altman, J. C. 1983. *Aborigines and mining royalties in the Northern Territory.* Canberra: Australian Institute of Aboriginal Studies.

Beaton, J. M. 1977. Dangerous harvest: Investigations in the late prehistoric occupation of upland south-east central Queensland. Ph.D. diss., Australian National University, Canberra.

————. 1981. Fire and water: Aspects of Australian Aboriginal management of Cycads. *Archaeology in Oceania* 17:51–67.

Bell, Diane. 1983. *Daughters of the Dreaming.* Melbourne: McPhee Gribble.

Berndt, R. M. 1982. Mining ventures: Alliances and oppositions. In *Aboriginal sites, rights and resource development.* Perth: University of West Australia Press.

Clark, Robin L. 1983. Pollen and charcoal evidence for the effects of Aboriginal burning on the vegetation of Australia. *Archaeology in Oceania* 18:32–37.

Coleman, J. A. 1982. A new look at the north coast: Fish traps and "Villages." In *Coastal archaeology in eastern Australia.* Canberra: Australian National University, Department of Prehistory.

Coombs, H. C. 1974. Decentralization trends among Aboriginal communities. *Search* 5:135–43.

Curthoys, Ann, and Andrew Markus, eds. 1978. *Who are our enemies?* Sydney: Hale and Iremonger.

Davidson, Bruce. 1980. The economics of pastoral and agricultural development in northern Australia. In *Northern Australia: Options and implications.* Canberra: Research School of Pacific Studies.

Dortch, Charles. 1979. Devil's Lair, an example of prolonged cave use in south-western Australia. *World Archaeology* 10:258–79.

Elkin, A. P. 1951. Reaction and interaction: A food gathering people and European settlement in Australia. *American Anthropologist* 53:164–86.

Evans, R. 1975. Keep white the strain. Introduction to *Exclusion, exploitation and extermination,* ed. R. Evans, K. Saunders, and K. Cronin. Sydney: Australian and New Zealand Book Company.

Flood, Josephine. 1983. *Archaeology of the Dreamtime.* Sydney and London: Collins.

Franklin, Margaret Ann. 1976. *Black and white Australians.* Melbourne: Heinemann Educational Australia.

Gould, Richard A. 1968. Preliminary report on excavations at Puntutjarpa rockshelter. *Archaeology and Physical Anthropology in Oceania* 3:161–85.

————. 1969. Subsistence behaviour among the western desert Aborigines of Australia. *Oceania* 39:253–74.

Hallam, Sylvia. 1975. *Fire and hearth.* Canberra: Australian Institute of Aboriginal Studies.

Hamilton, Annette. 1982. The unity of hunting-gathering societies: Reflections on economic forms and resource management. In *Resource managers: North American and Australian hunter-gatherers,* ed. Nancy M. Williams and Eugene S. Hunn. Boulder, Colo.: Westview Press.

————. 1984. Spoon-feeding the lizards: Culture and conflict in central Australia. *Meanjin* 43:363–78.

Hartwig, M. C. 1965. The progress of white settlement in the Alice Springs district and its effects upon the Aboriginal inhabitants, 1860–1894. Ph.D. diss., University of Adelaide.

Heatley, A. J. 1983. Problems and group power perceptions. In *Territorians or mobile Australians: A profile of the urban electorate*, ed. D. Jaensch and P. Loveday. Darwin: Australian National University, North Australian Research Unit.

Horton, D. R. 1982. The burning question: Aborigines, fire and Australian ecosystems. *Mankind* 13:237–51.

Howitt, Ritchie, and John Douglas. 1983. *Aborigines and mining companies in northern Australia*. Sydney: Alternative Publishing Company.

Jones, Rhys. 1969. Fire-stick farming. *Australian National History* 16:224–28.

Kimber, R. G. 1984. Resource use and management in central Australia. *Australian Aboriginal Studies* 2:12–23.

————. 1984. Aboriginal perspectives of the land and its resources. In *Aboriginal sites, rights and resource development*, ed. R. M. Berndt. Perth: University of Western Australia Press.

Latz, P. K., and G. T. Griffin. 1978. Changes in aboriginal land management in central Australia. In *The nutrition of Aborigines in relation to the ecosystem of central Australia*, ed. B. S. Hetzel and H. J. Frith. Melbourne: CSIRO.

Lewis, Henry T. 1982. Fire technology and resource management in aboriginal North America and Australia. In *Resource managers: North American and Australian hunter-gatherers*, ed. Nancy M. Williams and Eugene S. Hunn. Boulder, Colo.: Westview Press.

Lhanhupuy, Wesley. 1982. Aboriginal perspectives of the land and its resources. In *Aboriginal sites, rights and resource development*, ed. R. M. Berndt. Perth: University of Western Australia Press.

Lourandos, Harry. 1980. Change or stability? Hydraulics, hunter-gatherers and population in temperate Australia. *World Archaeology* 11:245–64.

Mackenzie, Ihian. 1980. European incursions and failures in northern Australia. In *Northern Australia: Options and implications*, ed. Rhys Jones. Canberra: Research School of Pacific Studies.

Maddock, Kenneth. 1981. Warlpiri land tenure: A test case in legal anthropology. *Oceania* 52:85–102.

Meehan, Betty, and Rhys Jones. 1980. The outstation movement and hints of a white backlash. In *Northern Australia: Options and implications*, ed. Rhys Jones. Canberra: Research School of Pacific Studies.

Meggitt, M. J. 1962. *Desert people: A study of the Walbiri Aborigines of central Australia*. Sydney: Angus and Robertson.

Morgan, H. M. 1982. The mining industry and Aborigines. In *Aboriginal sites, rights and resource development*, ed. R. M. Berndt. Perth: University of Western Australia Press.

Myers, Fred R. 1976. To have and to hold: A study of persistence and changes in Pintupi social life. Ph.D. diss., Bryn Mawr College.

———. 1982. Always ask: Resource use and land ownership among Pintupi Aborigines of the Australian western desert. In *Resource managers: North American and Australian hunter-gatherers*, ed. Nancy M. Williams and Eugene S. Hunn. Boulder, Colo.: Westview Press.

Nathan, Pam, and Dick Leichleitner Japanangka. 1983. *Settle down country*. Malmsbury, Vic.: Kibble Books and Central Australian Aboriginal Congress.

Newsome, Alan. 1980. Grazing versus wildlife in northern Australia: A note. In *Northern Australia: Options and implications*, ed. Rhys Jones. Canberra: Research School of Pacific Studies.

Peterson, Nicolas. 1972. Totemism yesterday: Sentiment and local organisation among the Australian Aborigines. *Man* 7:12–32.

———. 1975. Hunter-gatherer territoriality: The perspective from Australia. *American Anthropologist* 77:53–68.

Peterson, Nicolas, ed. 1981. *Aboriginal Land Rights: A handbook*. Canberra: Australian Institute of Aboriginal Studies.

Reynolds, Henry. 1988. *The Law of the Land*. Ringwood, Vic.: Penguin Books.

Rolls, Eric. 1969. *They all ran wild: The story of pests on the land in Australia*. Sydney: Angus and Robertson.

Rowley, Charles D. 1970. *The destruction of Aboriginal society*. Canberra: Australian National University Press.

Saddler, Hugh. 1980. Implications of the battle for the Alligator Rivers: Land use planning and environmental protection. In *Northern Australia: Options and implications*, ed. Rhys Jones. Canberra: Research School of Pacific Studies.

Smith, Len R. 1980. *The Aboriginal population of Australia*. Canberra: Australian National University Press.

Stanner, W. E. H. 1964. *On Aboriginal religion*. Oceania Monograph 11. Sydney: University of Sydney.

Strehlow, T. G. H. 1947. *Aranda traditions*. Melbourne: Melbourne University Press.

———. 1965. Culture, social structure, and environment in Aboriginal central Australia. In *Aboriginal man in Australia*, ed. Ronald M. and Catherine H. Berndt. Sydney: Angus and Robertson.

―――. 1970. Geography and the totemic landscape in central Australia: A functional study. In *Australian Aboriginal anthropology*, ed. Ronald M. Berndt. Nedlands: University of Western Australia Press.

―――. 1978. *Journey to Horseshoe Bend*. Sydney: Angus and Robertson.

Testart, A. n.d. La conservation des produits végétaux chez les chasseurs-ceuilleurs. In *La conservation des grains à long terme*, vol. 2. Marseilles: CNRS.

Tonkinson, Robert. 1978. *The Mardudjara Aborigines: Living the dream in Australia's desert*. New York: Holt, Rinehart and Winston.

Toyne, Phillip, and Daniel Vachon. 1984. *Growing up the Country*. Victoria: McPhee Gribble/Penguin Books.

Von Sturmer, John. 1982. Aborigines in the uranium industry: Toward self-management in the Alligator Rivers region. In *Aboriginal sites, rights and resource development*, ed. R. M. Berndt. Perth: University of Western Australia Press.

Wallace, Noel M. 1977. Pitjantjatjara decentralisation in northwest South Australia: Spiritual and psycho-social motivation. In *Aborigines and change: Australia in the '70s*, ed. R. M. Berndt. Australian Institute of Aboriginal Studies; Atlantic Highlands, N.J.: Humanities Press.

White, J. Peter, and James F. O'Connell. 1979. Australian prehistory: New aspects of antiquity. *Science* 203:21–28.

Chapter Nine

The Response of the Kenya Maasai to Changing Land Policies

Solomon Bekure and Ishmael Ole Pasha

INTRODUCTION

The history of how colonial settlers across the world dispossessed indigenous people of their best lands and pushed them into marginal areas is well known. Although the colonial days are over, land dispossession continues today in the name of resource development. The preceding chapters by Morris and Hamilton give ample evidence of this in North America, northern Europe, and Australia. Indigenous people did not easily give up their precious lands. They resisted and fought with all their might and wit but were in the end defeated by a combination of superior firepower and political maneuvering that often pitted indigenous ethnic groups against one another to the advantage of the colonial settlers. This chapter deals with the dispossession of one indigenous group in Africa, the Maasai of Kenya, and how they coped with it. Kenya's independence did not restore "tribal land rights" from the precolonial era; it fossilized the colonial dispossession.

The Maasai people, who currently inhabit southwestern Kenya and northwestern Tanzania, believe they originated from the area between Lake Turkana and the Dodos-Karamojong escarpment in Uganda. They gradually moved southward in the late fourteenth and early fifteenth centuries (Huntingford 1953). By the early seventeenth century they had occupied the Rift Valley as far south as the Serengeti plains in Northern Tanzania. By the time the Germans and the British colonized eastern Africa in the late nineteenth century, the Maasai had penetrated far south into the heart of present-day Tanzania, covering about 200,000 square kilometers of territory (map 9.1).

The Maasai were divided into tribal sections, run by an age-set system of political administration. Their armies comprised warrior groups known as *morans* that were taught warfare tactics as part of the cultural process of educating and inducting boys into adulthood. The *morans* were trained and disciplined as raiding and defense platoons. Until the early 1880s the Maasai were a formidable force in eastern Africa, controlling a vast territory. Although Jacobs (1979) disputes the description as exaggerated to suit British interests, Sir Charles Eliot, the first governor of Kenya, wrote: "They [the Maasai] successfully asserted themselves against the Arab slave traders, took tribute from all who passed through their country and treated other races, whether African or not, with the greatest arrogance" (Great Britain 1934).

Raising livestock played a prominent role in the life of the Maasai. Until recently, their diet was derived mainly from livestock and consisted of milk, meat, and blood (Jacobs 1972). Their livestock also served various purposes beyond the production of food. Hides and skins were used for shelter, clothing, and bedding. Slaughtering livestock was the principal means of celebrating weddings, circumcisions, and religious ceremonies. Livestock was the medium of exchange and payment of brideprice and was used for developing and cementing social ties and loyalties. Livestock provided the only means to store wealth and acquire the power and prestige concomitant with wealth. Thus the Maasai maintained large herds and flocks to satisfy these needs and to ensure their long-term survival in the face of recurrent droughts and epidemic diseases.

The expanse of territory they controlled was used for extensive rotational grazing. The well-watered high grounds with permanent water sources were grazed during the dry season, and the lowlands of the Rift Valley were grazed during the wet seasons. With access to vast grazing resources, both human and livestock populations grew rapidly. The Maasai were at the height of their territorial and political power in the 1880s (Huntingford 1953). A series of disasters during the 1890s resulted in great loss of human life and livestock, which diminished their wealth and dissipated their might. A rinderpest epidemic broke out about 1890 and was followed a year later by a serious drought and in the next year by a smallpox outbreak. The combined effect of these catastrophes claimed as much as 50 percent of the human pop-

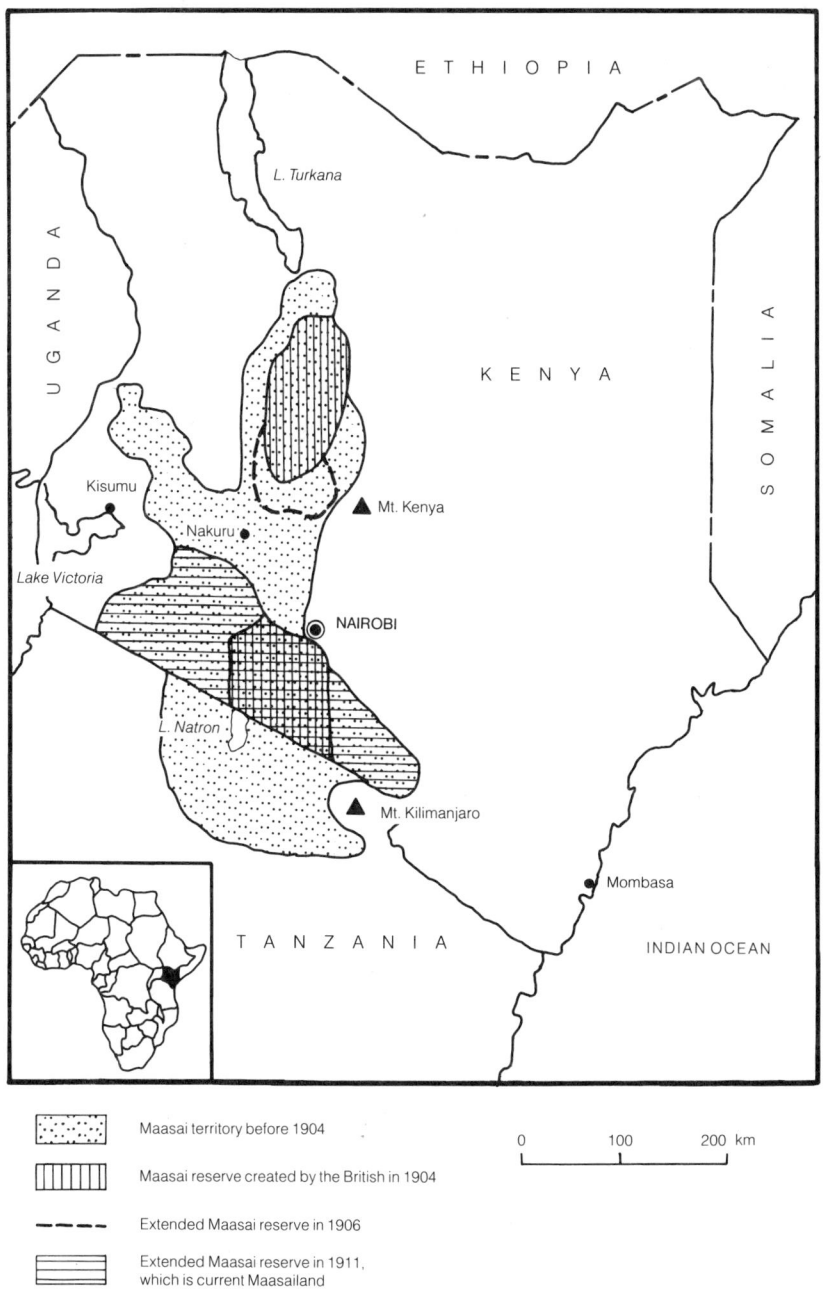

Map 9.1 Territory occupied by Maasai before and after dispossession during 1904 to 1912.

ulation and 80 percent of the livestock (Huntingford 1953). Just about that time Lenana and Sendeyu, the sons of the paramount chief of the Maasai, Laibon Mbatian, started a fratricidal war over succession to the leadership of the Maasai. At first Sendeyu managed to gain the upper hand and pushed Lenana and his supporters into Kikuyuland. The powerful Purko subtribe, who were at first neutral, threw their weight on Lenana's side. Sendeyu and his supporters, the Loita subtribe, were defeated and fled into Tanzania. This internal warfare resulted in extensive loss of life and further weakened the might of the Maasai (Great Britain 1934). It was precisely at this moment that the British incursion into eastern Africa started. In 1895 the Maasai in Kenya numbered only about 40,000, hardly a match for the British forces. The leader of the Maasai, Lenana, quickly acknowledged British rule and was spared military expeditions such as those sent against the Nandi, who resisted British incursion into their territory (Huntingford 1953).

BRITISH COLONIAL LAND POLICY IN MAASAILAND

European settlers were arriving in Kenya in increasing numbers, and by 1903 the colonial administration had received a large number of applications seeking permission to settle on land occupied by the Maasai in the Rift Valley. The issue of the time was how to accommodate the settlers' requests for land and at the same time be seen to uphold the rights of the Maasai. Sir Eliot succinctly summarized the rationale for the colonial policy that followed:

> The right of the Maasai to inhabit particular districts . . . is undoubted, . . . but their right to monopolize particular districts, and keep everybody else out . . . appears to me most questionable. As a matter of expediency it may sometimes be best to make reserves, but, as a matter of principle, I cannot admit that wandering tribes have a right to keep other and superior races out of large tracts merely because they have acquired the habit of straggling over far more land than they can utilize. (Great Britain 1934)

Thus the colonial administration made a policy decision to grant Maasai land in the Rift Valley to the European settlers. It created two Maasai reserves. The northern reserve was in Laikipia north and east of the Aberdares. The southern reserve extended from the Uaso Nyiro River in the west to the foothills of Mount Kilimanjaro in the east (map 9.1). The intention of the British colonial administration was to evacuate all Maasai from the Rift Valley into these reserves, which they could not leave without the explicit permission of the district commissioners.

The Maasai chiefs were summoned to a meeting at Naivasha on August 9, 1904, at which they were told of this decision and made to sign the "Maasai Agreement of 1904," which dispossessed them of their right to use about 4.6 million hectares of land. They were allowed access to only one-third of the territory they originally occupied. While some Maasai evacuated as required, others, particularly the Elburgo, Gakunki, Loita, Damat, and Laitutok, responded with passive resistance by not moving to the reserves. Other Maasai groups moved into areas far beyond the confines of the southern reserve as constituted in the agreement (Great Britain 1934). Those who moved to Laikipia, the northern reserve, complained bitterly of the inadequacy of the area for their livestock. They persuaded the British authorities to extend it in 1906 (map 9.1) (Huntingford 1953).

Two problems ensued in the administration of the two Maasai reserves. On the one hand, the British were faced with continued violation of the agreement by the Maasai grazing outside the reserves. On the other hand, the spatial splitting of the Maasai into two reserves eroded the power of Chief Lenana over the whole tribe. He presented his dissatisfaction with the physical separation of the tribe to the colonial administration and persuaded it to reunite all Maasai in a much larger southern reserve. The Maasai would evacuate the northern reserve, which would be opened for European settlement. A second agreement to this effect was signed on April 14, 1911. The southern reserve was extended to the east and the west by incorporating an additional 585,000 hectares of land. The "Maasai move" from the northern to the southern reserve started in June 1911. A congestion of livestock developed on the Mau Mountains, which resulted in some loss of cattle. The move was suspended when Maasai started to graze their livestock on land held by European settlers. It was recommenced a year later, at which time the leaders of the Maasai engaged a British lawyer to represent them in stopping the move and to allow those who had already moved to return to Laikipia. The lawyer brought a suit against the attorney general, but it was readily dismissed by the court. An appeal to the court of appeals for eastern Africa was similarly dismissed, the court stating that it had no jurisdiction over the matter. Thus the Maasai move continued until it was completed in March 1913 (Great Britain 1934). The Maasai registered their complaint that the land allocated to them in the southern reserve was insufficient to support their livestock. They managed to get further extensions into the Trans-Mara area (about 160,600 hectares), the Kilimanjaro foothills (about 966,600 hectares), and the Mau Mountains (about 85,000 hectares). The extension in the Mau area was to comply to the 1911 request of the Maasai to include the Mau in their reserve, since they believed it was their sacred "Promised Land." The 1913 extension of the southern reserve brought the land finally allocated to the exclusive use of the Maasai to 3.8 million hectares compared with 7.7 million before the reserve (Great Britain 1934). This meant that the Maasai were relegated to only half of the grazing area they had had access to. The dispossession is even greater considering they lost some of the better-watered areas.

Left in their reserve and aided by the colonial government's veterinary vaccination program to protect European settlers' cattle, the Maasai managed to rebuild their livestock numbers rapidly in the next two decades. Meanwhile other African tribes, particularly the Kikuyu, who were pushed into what were known as the "native" or "African" reserves, created to pave the way for European settlement of the high-potential agricultural areas, were becoming highly congested. This led to a considerable debate in the colonial administration on how so few Maasai could be allowed the exclusive use of a large expanse of land, some of which was well suited for crop cultivation. The Kenya Land Commission of 1932, set up to assess the land needs of the "native" population, saw the Maasai agreements of 1904 and 1911 as a major impediment to alleviating the congestion in the other native reserves. It argued, "The maldistribution of native population would be ameliorated if a less exclusive policy were adopted in the Maasai Reserve" (Great Britain 1934). It proposed that it would be in the best interests of the Maasai if the arable areas in the Maasai reserve were leased to other native tribes, subject to appropriate safeguards protecting Maasai interests. It even concluded,

> From these considerations we are satisfied that too rigid an adherence to the terms of the Agreement is not in the interest of the Maasai. When we turn from the interests of the Maasai and consider the interests of other tribes, it is

clear that the permanent entail of a vast area of land for the benefit of a tribe which makes little use of it and left to itself would certainly not be able to keep it, must appear unjust; especially when one, at least, of the neighbouring tribes is living in a state which borders on congestion. (Great Britain 1934)

The main cause of this congestion—dispossession from prime agricultural land that belonged to the natives—by the colonial government for the settlement of the Europeans was conveniently forgotten. The Kenya Land Commission's recommendation was to look into every possibility by which the colonial administration could legally amend the 1904 and 1911 Maasai agreements to further dispossess Maasai from the arable parts of their land for settlement by other native tribes in order to defuse the political tension and revolt that congestion in the other native reserves would eventually entail. Encouraged by official support and using relatives who were married to Maasai as an entry point, other African tribes, especially the Kikuyu, expanded cultivation into the areas around the Ngong Hills and the foothills of Kilimanjaro. These were important dry-season grazing areas for the Maasai. In the meantime Maasai livestock was increasing rapidly, resulting in overstocking and the deterioration of grazing land that was not infested by tsetse flies. The colonial government became concerned about the soil erosion and range degradation that would ensue and introduced measures that were supposed to stem these adverse effects. In the late 1930s poll taxes were imposed to force the Maasai to sell their livestock, but the Maasai response was unfavorable. During the Second World War the colonial administration was forced to increase its resources for the war effort and imposed a quota of 2,000 cattle a month to be sold by the Maasai. This managed to destock Maasai cattle by about 66,000 (about 7 percent of the cattle population each year) during 1943–46 (Campbell 1979).

Meanwhile the delineation and later gazetting of wildlife reserves and the unabated influx of other tribes to Maasai land further reduced the area available for grazing. The Maasai officially requested the colonial government to halt the migration of other tribes into Maasailand. The clearing of forests for cultivation was also causing the authorities some concern. A new regulation was issued in 1947 requiring other tribesmen to obtain permits for entering Maasailand. This, however, had little effect in curbing the influx. It was finally the Mau Mau insurgency that led to the emergency declaration in

1952 and stopped the process when the Kikuyu were repatriated to their reserve (Campbell 1979).

The large concentration of Maasai livestock in a much reduced area of land led to overstocking. The colonial government, keen to avoid degradation of the rangelands, introduced a scheme of forced destocking, which was fiercely resisted by the Maasai. The scheme was soon abandoned to avoid the discontent that was festering among the pastoralists (Migot-Adholla and Little 1980). The colonial government established the African Development Board (ALDEV), which was given the responsibility to eradicate the tsetse fly so as to open up new rangeland for human occupation and thus relieve the overstocking of the inhabited rangelands, to carry out vaccination against rinderpest, to drill boreholes, to construct dams, and to administer grazing schemes. The last were resisted, and ALDEV's efforts were too modest to have the desired effect. The Swynnerton plan (1954) for developing African agriculture noted that overstocking and uncontrolled grazing were rapidly converting the rangelands into near desert. It advocated launching livestock marketing schemes that would reduce the stocking rate and improve the quality of the remaining stock. This resulted in the creation of the African Livestock Marketing Organization (ALMO) (Migot-Adholla and Little 1980). However, the prices offered to the Maasai were so low that they refused to sell their cattle (Jacobs 1972). The cattle population in Maasailand reached a peak in 1960. There were an estimated 680,000 head of cattle in Kajiado district alone. A severe drought struck during 1960–61 that killed more than 60 percent of the livestock population (Campbell 1979).

POSTINDEPENDENCE LAND POLICY IN MAASAILAND

Following the 1960–61 drought and independence in 1963, concern over the development of the arid and semiarid areas became urgent. A Range Management Division was created in the Ministry of Agriculture, charged with advising the government and implementing programs for the conservation, management, and use of the rangelands. With the assistance of the United Nations Development Program (UNDP)/Food and Agriculture Organization (FAO), several detailed studies were conducted to prepare inventories of the range resources including human, livestock, and wildlife populations, soils, vegetation, and hydrology as a basis for preparing the Kenya Livestock Development Project (KLDP). The planners recognized four types of land use in the rangelands and set the following objectives for each type (Pratt 1968):

—The production of high quality beef for the export market would be encouraged in the European settler run commercial ranches (1 million ha);

—The transformation of subsistence pastoralism into commercial livestock production would be promoted in the pastoral rangelands (43.5 million ha) with consideration for preserving wildlife and rehabilitation of degraded areas;

—Economic benefits from tourism would be maximized by the proper management of National Parks and Game Reserves (2.8 million ha); and

—The unoccupied and unallocated lands would be opened up for commercial ranching while accommodating the future need of the people residing in their vicinities.

Security of tenure was advocated as a key instrument in promoting the development of the pastoral rangelands. Legislation was therefore enacted to allow the adjudication of land and the issuing of title deeds to groups of pastoralists (Land Group Representatives Act, 1968). It was believed that with security of tenure the tendency to overstock the ranges would diminish, the incentives for investing in range improvement would increase, and land titles could be used as collateral for obtaining loans for investment. It was hoped that these improvements would eventually lead to "a radical transformation of nomadic subsistence production into a sedentary, more commercially oriented system" (Von Kaufman 1976). Thus started the carving up of the large expanse of communal Maasailand into numerous small units of individual and group ranches. Although an emphasis was placed on establishing group ranches, land was also adjudicated to elite Maasai individuals, who by practicing improved husbandry measures were supposed to demonstrate to the group ranch members the benefits of modern range and livestock management.

The group ranch concept called for major changes in Maasai social organization and livestock-management strategies (Grandin 1981). The adjudication of land to registered members meant that they were excluded from using land outside the boundary of their own group ranch as a matter of right. Grazing quotas were supposed to be allocated to each member in order to match animal numbers to the carrying capacity of the ranch. When animal numbers exceeded the prescribed limit the group ranch committee, elected by the members of the ranch, would force reductions. Those holding livestock beyond their quota had to destock. The group ranch committee would oversee all communal (group ranch) affairs, which in essence replaced the traditional authority of the chiefs. Fifteen group ranches were established in the Kaputiei section of Maasailand during the first phase (1969–74) of the project, then a further thirty-seven were established during the second phase in Kajiado, Narok, and Samburu districts.

Although many Maasai did not fully grasp the ramifications of the group ranch approach, it appears that the desire for the security of land tenure (prevention of alienating land to Maasai and non-Maasai individuals) and the opportunities for the development of water sources and dips that the supporting project offered seemed to have coalesced to make Maasai receptive to the concept (Hedlund 1971). Over 80 percent of Maasailand has been registered and has been adjudicated mostly into group ranches and to a much lesser extent into individual ranches. The World Bank–assisted Kenya Livestock Development Project has financed infrastructural development and steer-purchase loans for both the group and individual ranches. Although the group ranches have in general shown reluctance to accept loans for these purposes, fifteen had been financed under phase 1 and an additional nine under phase 2 of the project as of the end of 1981.

The group ranch concept has had lasting and far-reaching effects on Maasai pastoralists—albeit some of them unintended or unforeseen. First, it has achieved the major goal for which it was accepted by the Maasai. The group ranches have helped to keep Maasailand in the hands of the Maasai by effectively preventing further incursions of other tribes in postindependence Kenya. That most of the land was adjudicated to groups who cannot sell it has prevented wholesale alienation of the land through the market. However, it has also provided the medium of transition from a communal form of land tenure to individual tenure, to which we shall return later.

Second, group ranches have fostered sedentarization among the Maasai, who have developed a sense of group ranch identity. The Maasai tend to graze their livestock inside their group ranches during years of normal rainfall. But since none of the group ranches are big enough to accommodate adequate dry-season grazing, in times of drought and poor or late rains the

Maasai are forced to seek grazing outside their group ranches. They use their social ties and relationships to gain access to grazing lands on other group ranches. Some better-informed Maasai deliberately registered their close relatives strategically in different group ranches to facilitate such access (Hedlund 1971). The Maasai recognize sedentarization as advantageous, especially with regard to their children's schooling. They can now have more material possessions, such as better housing and furniture, which nomadic life precluded (Grandin 1981).

Third, the Maasai have improved their management of livestock. Through the livestock development project that accompanied the adjudication of group ranches, new water sources were developed and dips constructed. The veterinary and extension services were active in the initial stages of the project, and the Maasai are now able to water their animals more frequently, use acaricides to control ticks that transmit diseases, administer drugs to sick animals, and provide salt licks, particularly to small stock. Those in group ranches that receive higher rainfall (500–700 millimeters) have been able to introduce large breeds, particularly Sahiwals. They have also learned that steers can be bought, fattened on the range, and sold for profit, though this was demonstrated to them in a rather perverse way. The development of water sources and dips on the group ranches was financed through loans provided by the Agricultural Finance Corporation (AFC). The loan was supposed to be paid back by the group ranch members in accordance with their stock (grazing) quota, which was never implemented. Furthermore, since in the old days the colonial government had developed boreholes at no cost to the Maasai, the ranch members claimed they believed those were likewise provided as grants. To make matters worse, the 1974–76 drought forced most of the Maasai to evacuate the ranches and travel far in search of grazing. Some of the engines and pumps on the group ranches were left unattended. When conditions improved and the Maasai returned, they found the engines and pumps had been stolen. So when the AFC was pressing for loan repayment, the Maasai became even more reluctant to pay for what they no longer had. The AFC devised an ingenious method of getting back its money. It bought immature steers and placed them on the group ranches under the care of the group ranch committee. The steers were grazed for up to one year and sold, and the AFC kept the profit as a partial payment, repeating this operation until the loan was paid off. Having witnessed such a practice, the Maasai would like to adopt it for themselves, and this is adding fuel to the pressure for individual tenure that provides collateral for securing loans.

Although we have listed the positive effects of group ranches, there are areas in which the intended consequences have not materialized or unforeseen or unstated consequences have developed. A major rationale and motivation for the group ranch concept was range conservation. Planners felt that the "tragedy of the commons" was operating in pastoral areas as each herd owner was maximizing herd size in unregulated competition with other pastoralists. By giving a small group exclusive title to smaller portions of land, they hoped that conservationist measures would be more readily adopted. This was felt to be particularly important lest providing increased watering facilities lead to even greater degradation.

Group ranch members were collectively meant to limit their aggregate livestock holdings to a level technically assessed as commensurate with the long-term natural resources of the ranch. This has not been achieved; as far as one can see, it has not even been tried, since it was difficult to convince the Maasai that it should be. The considerable wealth differential among Maasai makes simple percentage reduction unacceptable. There is as yet little evidence that planners have determined what number and combination of animals will meet a family's food needs from year to year. As Dyson-Hudson (1982) writes, "General and voluntary income redistribution is no more feasible among the Maasai than it would be in most other producers."

Whether a long-term degradation of the rangelands is taking place is very difficult to ascertain. Examination of range transects by Skovlin (1980) and Njoka (1984) ten to twenty years after they were established was inconclusive. Some transects showed decreases in perennial grasses, some showed increases, and others remained about the same. The only area where Skovlin reported severe degradation of the range was in portions of Samburu district. What is well known is that Maasai population has increased rapidly over the past three decades at about 3 percent a year, reducing their per capita cattle holdings from thirteen head in 1950 to about five head in 1980 (Grandin 1988). Though their livestock numbers have fluctuated tremendously owing to decimation by recurring droughts (once in eight to twelve years), the long-term trend has been a rising one that has increased the pressure on range resources. The Maasai no longer use burning as a tool of grazing management, because there is not enough standing grass to yield a hot burn even in years of excellent rainfall. The development of new water points and the availability of trypanocidal drugs and acaricides have opened up hitherto inhospitable outback grazing areas, slightly reducing the pressure on the range as a whole. The Maasai have also diversified their livestock enterprise by increasing the number of sheep and goats they keep. This small stock, especially goats, browses different species of plants than cattle, thus giving the Maasai complementary livestock production without proportionately taxing the range. The Maasai have also changed their diet by gradually increasing the amount of maize they eat, enabling them to live off fewer animals per capita than three decades earlier. So far all these adjustments, coupled with the occurrence of drought, have tended to limit the potential damage of the range by rising livestock numbers. However, the scope for such adjustment in the future will be limited, with adverse consequences on the range.

Traditionally, Maasai authority is vested in the elders. As a major departure from this, the group ranch constituted a new social formation for the Maasai, involving "an alien political concept of decision-making and enforcement by a committee of elected representatives" (Grandin 1981). It required a committee of ten persons to manage the affairs of the group ranches. This called for making binding and prompt decisions about shared natural resources, individual livestock holdings, the development of resources, the management of ranch properties, and the servicing of collective debt.

> This the group ranch committee generally could not do. Nothing in their previous decision-making experience, in their cultural values, nor in the existing production organization prepared them to reach—let alone enforce—such binding commitments. Decision-making in traditional pastoral systems is based on decision-avoidance until the point where the options are so few and the reaction so necessary that voluntary collective response is assured. Attempts to force a decision before that point of general perception simply lead to individual producers (who are basically autonomous) breaking away and seeking solutions on their own. There is thus a tendency for the committee not to meet; or if it meets to deal in non-controversial generalities; or if it addresses specifics, to be unable to reach a conclusion; or if it reaches a conclusion to be unable to enforce it. (Dyson-Hudson 1982)

The ranch committee is assumed to represent the collective interests of the producers, who are the ranch members. The actual situation is more complex because the committee members represent variable ties of age set and clan within the ranch, are individually subject to age set, clanship, and friendship pressures from outside the ranch, and are variably subject to regional and national political pressures according to their own beliefs and ambitions. These last two points, taken together, mean that there are plenty of reasons for disagreement and few organizational options for resolving them (Doherty 1971). For these and other reasons, the group ranch committees have in general been ineffective in discharging their duties and responsibilities. They have been unable to properly manage and maintain dips, water pumps, and engines. They have failed to allocate and enforce stock quotas. They have not attempted to organize and control grazing patterns effectively, nor have they managed to enforce the group ranch boundaries. They have been unable to collect repayment of the AFC loans. In short, they have failed to manage the affairs of the group ranches in the manner envisaged by the planners.

> Some of the problems found on group ranches now are attributed to the fact that the close communication with the Maasai and supervision of group ranches (envisaged in the project) never materialized. There were staffing problems at AFC which in any case was not experienced in dealing with traditional pastoralists. For most of the life of the project the office of the Registrar of Group Representatives was staffed by only one senior person, a completely inadequate provision for the task of supervision. No group ranch has had a qualified manager. (Grandin 1981)

Although the Range Management Division had qualified staff to provide technical information for planning, they were ill equipped to give extension advice on how to run group ranch affairs. Senior elders on group ranch committees claimed that extension officers sent to work with them were too young, lacked a pastoral background, did not speak the Maa language, and had nothing new to teach them. Thus there was little dialogue and therefore little scope for assistance. Furthermore, committee members, especially the chairmen, spent a lot of time and also money on tending to group ranch matters for which they received no compensation (Grandin 1981). This hampered their active participation. Membership of group ranches has been limited to those registered originally. This has had negative effects on the quality of committee membership. The young, educated, and ambitious Maasai who want to participate in the decision making with respect to group ranch affairs have been barred by the older age set of their fathers because they were not registered as members, and the issue has generated conflicts between the young and old age sets, leading to an intense push for the subdivision of group ranches.

SUBDIVISION OF GROUP RANCHES

An unintended or unstated consequence of the implementation of the group ranch concept is the subdivision of communal grazing lands into individual plots. The pressure for such subdivision has been building up in Maasailand in recent years. Fierce debate has been going on between those who call themselves progressives and conservatives, between the young and the elderly, and between politicians and extension officers. The main factors that contribute to the pressure for subdivision are the general policy of the government of Kenya to adjudicate land to individual Kenyans on a freehold basis; the agitation of the young Maasai, who have been excluded from being legally registered as members of the group ranches; the value of freehold land title deeds as collateral for securing loans as well as for the outright selling of land; and the appalling inefficiency with which group ranch affairs are being conducted by the committees.

The official policy of the government of Kenya since independence has been to pursue vigorously the adjudication of land to Kenyans based on freehold tenure. This was done swiftly in the high-potential areas through the program of land settlement and land transfer in the former scheduled areas owned by the white settlers. By 1970, about 1.2 million hectares of land was adjudicated in the high-potential areas in contrast to only 0.21 million in the range areas, including individual farms, ranches, and group ranches. The 1974–78 development plan stated: "The land adjudication and registration programme is now larger than ever before and this expansion has been regarded as a prerequisite for faster agricultural development. The security of title which it provides makes farmers more willing to undertake long term farm improvements, while it enables them to obtain agricultural credit more easily through pledging their land as security" (Kenya 1974). Realizing this very fact, a number of elite Maasai persuaded the county councils to adjudi-cate them individual ranches with freehold title. The argument advanced was that since the individual Maasai ranchers were educated, they would run their ranches using modern techniques: watering, dipping, and vaccinating their stock regularly, practicing rotational grazing, introducing improved breeds (Boran and Sahiwal), and fattening steers and selling them at youn-ger ages. Thus the individual ranchers, being of the same ethnicity, could show their traditional brothers in the group ranches what could be achieved given certain differences of outlook, capital availability, and learnable skills. Most of those Maasai who own individual ranches in Kajiado district are re-garded as a wealthy class. Most have permanent stone houses, some own motor cars, they own business premises in trading centers, and above all they have large herds of Sahiwal cattle (a breed of cattle believed to be supe-rior to the African Zebu). Many group ranch members, particularly the young, believe that if they too owned individual ranches they could have easy access to AFC loans and that their standards of life would be nearly the same as that of the individual ranchers. Furthermore, in the light of the frus-trations of operating group ranches, they seem intent on realizing the au-tonomy that characterizes the herd owner in the traditional system but is curtailed in the group ranches. They are thus convinced that subdivision of group ranches would enhance their self-advancement, and they actively pur-sue this goal. This is reinforced by official government policy, which encour-ages private ownership, and by pronouncements of politicians who espouse the same.

There is a major difference between the amount of land that was allocated to individual ranches in the past (about 800 hectares each) and what can be allocated now. If all the group ranches subdivided their land on the basis of equal shares to each member, each member would receive on the average about 100 hectares (A. H. Jacobs, pers. comm.). At the time of writing, 29 of the 52 group ranches in Kajiado district have passed resolutions to subdivide their group ranches into individual plots (A. H. Jacobs, pers. comm.). Seven of these have gone ahead and subdivided the land equally among the registered members but are awaiting the official adjudication and issuance of title deeds by the government (Pasha 1986). The remaining twenty-two are at various stages in the process leading to subdivision and adjudication. Some group ranches, although their members have passed a resolution to subdivide, are not actively implementing it but are watching what is happening to those who are actively engaged in the subdivision. One hopes they will learn from the pitfalls and mistakes of their predecessors. Others are arguing about how the land should be subdivided. Central to this is the issue of equity. The poor members of the group ranch and those with few children and livestock want equal subdivision among all members, while those with large households and greater number of livestock prefer division either by household or by livestock numbers.

A closer look at the seven group ranches that have already implemented subdivision shows that proximity to urban centers, availability of arable and irrigable land, and long experience with group ranching seem to have been an added influence in the decision to subdivide. Proximity to urban centers and availability of arable or irrigable land increase the value and marketability of freehold land, making subdivision more attractive. A common feature of those ranches that have resolved not to subdivide seems to be the lack of arable land. All of these are in the drier part of the western, southern, and southeastern parts of the districts. The only exception to this is Kimana group ranch, which has patches of irrigable land along the Kimana swamps.

Some Maasai are opposed to the subdivision of group ranches, for four reasons: The first is the alienation of land to non-Maasai. They believe that other ethnic groups that were warded off by the group ranches will now find it very easy to buy pieces of Maasailand, and therefore they fear an influx of outsiders taking up especially the arable parts. Second, they fear that this will lead to cultivation of large tracts of land, which will result in severe soil erosion, as happened in other parts of Kenya such as in Machakos district. Third, they fear that an influx of non-Maasai settling among them will result in the erosion and eventually the total loss of Maasai culture, which they want preserved. Finally, they see a danger that once a subdivision of group ranches has been effected, individual owners will tend to protect their private property much more. The usual livestock movements across the group ranch will thus be curtailed, especially by those who might undertake cropping, since they will be forced to fence their farms or gardens to protect the crops from either wildlife or livestock. With such new development the usual free movement of livestock on the ranch will be further constrained.

CONCLUSION

Land-use policy in Maasailand has undergone major changes in the past one hundred years. The Maasai were dispossessed of half of their land, including their best dry-season grazing areas, first by British colonialism, later by the influx of other African tribes and by the establishment of wildlife reserves and game parks. The Maasai at first responded with passive resistance, since they were powerless to resist with violence. They have not been and will not be able to regain these lands. They have had to cope with a declining resource base in spite of their increasing human population. The negative effects of adjusting to this situation have been somewhat lessened by opening up inhospitable areas through veterinary care and water technology and increasing the proportion of cereals in their diet, thus reducing the per capita dependence on livestock. The postindependence policy of the Kenya government to adjudicate land to individuals and small groups of producers—while preserving the remains of Maasailand in the hands of the Maasai—has encouraged the subdivision of communal grazing land into individually owned plots.

If such a subdivision is implemented, it will have important implications for the way the Maasai manage their livestock. The rich with very large number of livestock will have to either buy or rent land. Otherwise they will not have sufficient grazing unless they sell their animals. Those who have too few cattle will rent out or sell their grazing land, buy more livestock, or both. Those who sell, by default or unwittingly, may find themselves displaced and unemployed. Subdivision will also have important implications for how hitherto communal resources are to be shared. It is essential that government not impose a solution on the Maasai. It should take an active role in assisting them to work out new mechanisms that will lessen the pain of adjusting to the new land tenure transformation.

REFERENCES

Campbell, D. J. 1979. *Development or decline: Resource, land use and population growth in Kajiado district*. Working Paper 352. Nairobi: Institute for Development Studies, University of Nairobi.

Doherty, D. 1971. *Factors inhibiting economic development on Rotian Olmakongo group ranch*. Working Paper 356. Nairobi: Institute of Development Studies, University of Nairobi.

Dyson-Hudson, N. 1982. Changing production strategies in east African range livestock systems: An overview. International Livestock Centre for Africa document prepared for the Programme Committee.

Galaty, J. 1980. The Maasai group ranch: Politics and development in an African pastoral society. In *When nomads settle*, ed. J. Galaty. Montreal: McGill University.

Grandin, B. E. 1981. *Report for the IBRD: Impact study of Kenya Livestock Development Programme Phase I*. Nairobi: International Livestock Centre for Africa.

———. 1988. Kajiado Maasailand: The socio-historical context and group ranches. In *Maasai herding: An investigation of pastoral production on group ranches in Kenya*, ed. S. Bekure, P. N. de Leeuw, and B. E. Grandin. Nairobi: International Livestock Centre for Africa.

Great Britain. 1934. *Report of the Kenya Land Commission* (September 1933). Cmd. 4556. London: His Majesty's Stationery Office.

Hedlund, H. G. 1971. *The impact of group ranches on pastoral society*. Staff Paper 100. Nairobi: Institute for Development Studies, University of Nairobi.

Huntingford, G. W. B. 1953. *The southern Nilo-Hamites*. London: International African Institute.

Jacobs, A. H. 1972. The pastoral Maasai of Kenya and Tanzania. In *Cultural source materials for population planning in East Africa*, vol. 2, *Innovations and Communication*, ed. Angela Molnos, 334–48. University of Nairobi.

———. 1979. Maasai inter-tribal relations: Belligerent herdsmen or peaceable pastoralists? In *Warfare among East African herders*, ed. K. Fukui and D. Turton. Senri Ethnological Studies 3. Japan.

Kenya, Republic of. 1974. *Development plan, 1974–1978*. Nairobi: Government Printers.

Migot-Adholla, S. E., and P. D. Little. 1981. Evolution of policy toward the development of pastoral areas in Kenya. In *The Future of the pastoral peoples: Proceedings of a conference held in Nairobi 4–8 August, 1980*, ed. J. G. Galaty, D. Aronson, P. C. Salzman, and A. Chouinard. IDRC-175e. Ottawa: International Development Centre.

Njoka, T. J. 1984. Evaluation of seasonal and drought effects on rangeland analysis in Kajiado district, Kenya. In *Proceedings of the workshop on land evaluation for extensive grazing (LEEG)*, ed. W. Siderius. Wageningen: International Institute for Land Reclamation and Improvement.

Pratt, D. J. 1968. Rangeland development in Kenya. *Annals of Arid Zone* 7(2).

Skovlin, J. M. 1980. An evaluation of ten-year rangeland changes at selected sites throughout Kenya. Report submitted to Head, Range Management Branch, Ministry of Livestock Development, Kilimo House, Nairobi, Kenya.

Von Kaufmannn R. 1976. The development of the range land areas. In *Agricultural development in Kenya: An economic assessment*, ed. J. Heyer et al. London: Oxford University Press.

Part Four

Indigenous Religion in
the Struggle for Land

🌿

The essays of this section, by J. Baird Callicott and O. Douglas Schwarz, discuss two questions: Can one discover an environmental ethic in indigenous non-Western religions and, if so, by what methods? Can the conservationist positions of indigenous religions be communicated to European-based cultures?

Ever since Lynn White published his "Historical Roots of Our Ecologic Crisis," on the impact of Christianity on environmental attitudes (arguing that the biblical ethic to "be fruitful, multiply, replenish the earth, . . . subdue it" gave Western humankind a license to exploit the natural environment), the role of religion and public ethical systems in setting conservation policy has been a matter of scholarly controversy (White 1967). White's thesis has been attacked by many scholars and by the poet Wendell Berry. Certainly in the works of Francis Bacon, the prophet of applied science, the notion of radically manipulating nature arises partly in opposition to orthodox conceptions of a static, unalterable nature and a divine wisdom that orders things in fixed, mathematical proportions. In consequence, many seventeenth-century thinkers regarded Bacon not as the traditional conservative Anglican he professed to be but as a "freethinker" (cf. Black 1970). Other scholars have attributed the supposed Western lack of a conservation ethic to Western religion's assumption of a God above nature as opposed to one within. The related notion of Manifest Destiny, based both on the

Western epic notion of destiny and on the post-Augustinian concept of a Christian responsibility to conquer other cultures, convert them, and impose on them the praxis of Europe, may have a role both in the creation of colonialism and in the failure of colonizing cultures to learn more about thrift and providence from those they conquered. In any case, few scholars have challenged the notion of a radical manipulability in nature, and colonialism based on that notion is European and Euro-American in origin.

As the modern environmental movement arose, scholars and advocates for native peoples both attributed to American Indians and other native peoples a land wisdom embedded in their ethical systems, their religions, or their rituals of hunting and farming. Shortly thereafter, Calvin Martin attacked this notion as it applied to recent centuries in his *Keepers of the Game* (Martin 1978), positing a kind of war between the animals and the Northeastern and Plains hunters who sought to harvest ever greater amounts of fur from their areas. Martin further argued that Indian religions are so incompatible with Christianity that they are incapable of having an effect on Western man.

Since Martin has become the center of the discussion over indigenous religions and the environment and the role of such religions in influencing the West, the Callicott and Schwarz essays below take primary issue with his theses that no land ethic can be extracted from Plains Indian religions and that they are so incompatible with Christianity that they can have no effect on Western people. Together the two essays assess the force of Native American conservation. The assessment of the relation between belief systems and behavior is still a controversial area—one that will, one suspects, be settled only through detailed examination of the use of story, ritual, and precept in the education of young people. However, if religious systems and mythoi are what they claim to be, the encoders of final truth for those who believe in them, the student of the relationship between high-energy and low-energy cultures cannot ignore their specific and differentiated effects on how people think about the world. The essays that follow endeavor to do this.

REFERENCES

Black, John N. 1970. *The dominion of man: The search for ecological responsibility.* Edinburgh: University Press.

Martin, Calvin. 1978. *Keepers of the game: Indian-animal relationships and the fur trade.* Berkeley: University of California Press.

White, Lynn. 1967. The historical roots of our ecologic crisis. *Science* 155:1203–7.

Chapter Ten

American Indian
Land Wisdom

J. Baird Callicott

The dialectics of contemporary popular mythology are such that a backlash against the image of traditional American Indians as native environmentalists may be expected to occur sooner or later as a matter of course. It will likely be sooner if Rudolf Kaiser's exposé of the spurious origins of what is surely the most celebrated example of American Indian land wisdom—the famous oration of Chief Seattle during the Point Elliot treaty negotiations of 1854–55 (Kaiser 1987)—becomes widely known in this country. Seattle almost certainly uttered an oration in his native language during these events. His speech was reconstructed in English, we cannot know how accurately, by Henry A. Smith, M.D., who was present (Starnes 1968, 34–37, 60–64). But the version of the speech (or, as it is sometimes called, letter) of Chief Seattle (to President Franklin Pierce), which has in recent years been quoted so ubiquitously, is not even the Smith rendition of unknown veracity: "I have seen a thousand rotting buffalos on the prairie, left by the white man who shot them from a passing train. I am a savage and do not understand how the smoking iron horse can be more important than the buffalo that we kill only to stay alive. . . . If we sell you our land, love it as we've loved it. Care for it as we've cared for it. . . . How can you buy or sell the sky, the warmth of the land? . . . The idea is strange to us" (Kaiser 1987, appendix 3). These familiar words are virtually household American Indian environmental rhetoric, but they, along with the other ecological and environmental pieties attributed to Seattle, were actually composed by one Ted Perry as a film script for a movie called *Home* produced by the Southern Baptist Convention in 1971–72, and interpolated into William Arrowsmith's redaction of Henry A. Smith's nineteenth-century rendition of Seattle's words—whatever they may have been.

Perry's "forgery" epitomizes the way traditional American Indian cultures have come to symbolize a lost but not forgotten harmony of human beings with nature in the wake of the popularization of ecology and public awareness of the environmental crisis. This latest American Indian icon expresses regret and outrage for the spoliation of a biotically diverse continent, but it also expresses hope that contemporary Euro-American society will emulate the ideal of a fitting human-nature relationship represented by traditional Native American peoples.

Ted Perry did not so much invent as embellish an image in the contemporary consciousness. And it is as difficult to specify precisely as to corroborate historically anything so diffuse and fatherless. A putative American Indian "land wisdom"—which may serve as a kind of generic term—resolves, upon closer scrutiny, into several distinct and apparently disparate types. Since there is, so to speak, a documentary horizon at roughly 1492 beyond which we cannot directly look, claims that one sort of environmental wisdom or another prevailed among traditional American Indian peoples are difficult to sift and verify. On the one hand, the manifest biotic capital of North America at first European landfall and, on the other, the archaeological evidence of occasional wastage and the extinction of many species of large mammals provide ambiguous indications of traditional American Indian environmental *behavior*. Moreover, the connection between cognitive culture and cultural behavior is complex and tenuous. Cultural ideals guide and inspire personal and collective behavior; they do not determine it. Conversely, episodes of behavior in violation of cultural norms do not necessarily invalidate or impugn them. Some method, or triangulation by means of several methods, is needed to assay more directly the environmental *beliefs*, *attitudes*, and *values* existing among traditional American Indian peoples.

Perry's elaboration of Seattle's speech is only one particularly sordid example of the casual way in which the idea of a traditional American Indian land wisdom has been propagated in contemporary culture. The critics of that idea have been equally irresponsible. For example, Daniel Guthrie in an especially scurrilous article published in *BioScience* cited, among other equally irrelevant circumstances, the environmental and ecological conditions on present-day Indian reservations as evidence that precontact American Indian peoples (who in his view represent "primitive man") were as environmentally and ecologically insensitive as their Euro-American successors (Guthrie 1971). In general, arguments for and against an American Indian land wisdom have been so distanced from specific cultural materials as to be virtually a priori in character. Nor has it even been at all clear exactly what is at issue.

Four distinct types of American Indian land wisdom have been proffered in the literature: utilitarian conservation, religious reverence, ecological awareness, and environmental ethics. Of these, environmental ethics best characterizes the attitude toward nature exemplified by the traditional Ojibwas, with whose worldview I am best acquainted. The Ojibwas are especially interesting for a variety of reasons, not least because they occupied both woodland and Plains environments. A similar, though structurally different, worldview, which would support and imply an environmental ethic, may also be found among the Sioux, a largely Plains people after the European invasion of North America (though I shall have less to say about it here).

In the 1930s William C. McCleod and Frank G. Speck attempted to persuade the literate public that principles of conservation were practiced among traditional American Indian peoples (McCleod 1936; Speck 1938, 1939). To my knowledge theirs are the earliest, and still among the best-informed, arguments for an American Indian environmental wisdom. By "conservation" Speck and McCleod expressly meant utilitarian conservation as it was understood and implemented in the 1930s and as it had been previously defined by Gifford Pinchot—the rational, prudent exploitation of natural resources to obtain from them the maximum sustained yield. Indeed, Speck explicitly employed evidence of conservation among American Indians to reach a more general conclusion, namely, that the then-common racist stereotype of traditional American Indians as peoples languishing at a subrational level of consciousness was erroneous. In other words, aboriginal conservation—the rational use of resources, Speck averred—proved that traditional American Indians were eo ipso rational.

Speck and McCleod based their conclusions on ethnographic observations of postcontact American Indians living in relative isolation from Euro-American civilization and practicing an apparently traditional way of life. Both Speck and McCleod relied heavily upon the *contemporaneous* institution of the family hunting territory and its division into annually rotated quarters—to permit the recovery of populations of game animals—as incontrovertible evidence of conservation among *aboriginal* northern woodland hunting-gathering peoples. Such an institution certainly suggests a careful, quantitative, empirical census of game resources and a management plan for their systematic exploitation—the basic elements of utilitarian conservation principles.

McCleod, however, is expressly aware that some question might be raised about the provenance of the family hunting territory system—that it might have been a response to the European fur trade and thus also ancillary to European economic concepts of natural resources, private property, supply, demand, and so on, or even that conservation was expressly taught to the Indians by whites. Still he insists that "there seems to be no doubt of its aboriginality" (McCleod 1936, 564). Subsequent research, of course, inclines to the opposite opinion—that the family hunting territory, with its cognitive correlates, was a postcontact development (Leacock 1954; Bishop 1970).

The distinctly cognitive, as opposed to behavioral, evidence for conservation marshaled by McCleod and Speck actually points to a second, completely different sort of environmental wisdom—which they identify as a "religious" reverence for nature and nonhuman natural entities. Speck rehearses miscellaneous examples of ceremonials, legends, and ritual practices among Algonkian hunter-gatherers and sums up his observations with this famous remark: "With these people no act of this sort is profane, hunting is not a war upon the animals, not a slaughter for food or profit, but a *holy occupation*" (Speck 1938, 260). And in a decidedly less dramatic style McCleod notes that "spiritual motives may have played a large part in giving rise to conservation. To primitive man there is soul in all things animate and (to us) inanimate. . . . [Thus] they appear to have rationalized [*sic*] their rude conservation in spiritual terms" (McCleod 1936, 562).

As McCleod, more clearly than Speck, seems to realize, "conservation" of "natural resources" may have been more a side effect of animism and religious veneration than a consciously envisioned goal. According to Speck and McCleod, waste or immoderate harvest of plants and animals was offensive to their spirits or spirit wardens. Hence without explicitly intending to conserve resources—their intent being rather not to offend nature spirits—the woodland hunter-gatherers Speck described might have achieved the same results adventitiously.

Indeed, it seems that the complex of concepts constituting deliberate utilitarian conservation would be cognitively dissonant with an alleged religious veneration of the environment. Speck concisely sums up the latter as follows: "The hunter's virtue lies in respecting the souls of the animals necessarily killed, in treating their remains in prescribed manner and in particular, in making use of as much of the carcass as is possible. These observances constitute religious obedience. The animals slain under the proper conditions and treated with the consideration due them return to life again and again. They furthermore indicate their whereabouts to the 'good' hunter in dreams resigning themselves to his weapons in a free spirit of self-sacrifice" (Speck 1939, 23).

To look upon animal and plant persons as impersonal material resources, to observe the details of their distribution, growth, recruitment, breeding habits and seasons, and so on with an eye to harvesting as many as prudently possible—that is, to coldly calculate the maximum or even optimum sustained yield of these "resources"—seems in and of itself to represent a dangerously irreverent attitude toward their spirits. Further, as more a matter of logic than likelihood, to believe that beaver, moose, and other slain animals returned to life upon proper disposal of their skeletons would seriously confound the calculations that are an essential element in the conservation cognitive complex.

Speck attributes to the American Indians he studied yet a third species of land wisdom that is more ideologically consonant with a religious veneration of nonhuman natural entities than utilitarian conservation is, but which—because it may exist apart from a religious framework of belief and practice—represents a distinct type. He says, "The idea of treating animals and plants after the principle of the *Golden Rule* was so typical that they literally used their privilege of hunting, fishing, and collecting, as though they were treating animals and plants with *the same consideration offered to human beings*" (Speck 1939, 23). Speck here attributes to woodland hunter-gatherers an ethical attitude—as distinct from an attitude of religious reverence—toward nonhuman natural entities. An ethical attitude toward certain beings may exist in the absence of worship, veneration, or reverence for them. We do not worship, venerate, or reverence other people, though we may respect them. As modern secular ethics illustrate, ethical attitudes may be divorced altogether from a religious or supernatural system of belief. Irving Hallowell's extensive, and extremely sympathetic and reflective, work among the Ojibwas, an Algonkian people, provides ample evidence that the Ojibwas regarded animals, plants, and assorted other natural things and phenomena as persons with whom it was possible to enter into complex social intercourse. And social interaction is, as it were, a matrix from which morality is forthcoming and in which it is in the first place exercised. (I shall comment more fully on Hallowell's observations and analyses as they bear on American Indian land wisdom in the next section).

A fourth distinct kind of land wisdom, recently the one most commonly attributed to pre-Columbian American Indians, is ecology. According to Stewart Udall (1972, 2–12)—who also attributes an environmental ethic to American Indians—an "ecological awareness" inhibited their potential despoiling of the environment. J. Donald Hughes (1977), Terrence Grieder (1970), G. Reichel-Dolmatoff (1976), William A. Richie (1955–56), Thomas W. Overholt (1979), and many other commentators also make essentially the same point.

The term "ecology," of course, has become more and more diluted as it has become more popular. It is often now used most loosely to mean roughly the same thing as "natural environment"—as in the sentence, "Strip mining is bad for the ecology." Correlatively, anyone who strongly advocates environmental protection may be called an ecologist, no matter how limited his or her training or even literacy in biology. Ecology, sensu stricto, is a subdiscipline of biology, and an ecologist is someone with an advanced degree (or at least a strong track record of professionally respectable amateur research) in that field. Thus to say that American Indians were native/intuitive/natural/original ecologists could mean, most loosely, that they were strong advocates of environmental protection or, most strictly, that they had advanced degrees in a subdiscipline of biology. The former notion is a historical/political anachronism, and the latter is absurd.

More generously interpreted, those who claim that American Indians possessed an ecological awareness may be taken to mean that American Indians had a thorough and systematic knowledge of the biota of their environments and, more especially, of the dynamic interactions, dependencies, and relationships among the several constituents of their environments. Such sophisticated knowledge is not at all unlikely to have accumulated among peoples living in an intimate, direct, and dependent relationship with nature. And that such knowledge actually exists among contemporary American Indians living in traditional lifeways is amply supported by ethnographic investigation (Nelson 1982, 211–28).

Not all contemporary scientific ecologists are environmentalists. Just as modern ecology may be put to use for strictly economic ends, so may native ecology have been. From the existence of an ecological awareness among American Indians, nothing follows about their *attitudes* and *values* respecting ecosystems.

These four basic types of putative environmental wisdom prevailing among traditional American Indians—conservation, religious reverence, environmental ethics, and ecological awareness—are very often confused or conflated by a single author in a single discussion. J. Donald Hughes, for example, promiscuously attributed all four forms of environmental wisdom (as well as feminism) to northeastern hunter-gatherers in "Forest Indians: The Holy Occupation" (1977).

The means of substantiating the notion that a land wisdom of one sort or another existed among traditional American Indians are equally varied and uncritical. The more popular discussions pay little attention to cultural specificity and rely primarily on what might be called testimonials for evidence or support. The previously mentioned (apocryphal) Chief Seattle, Chief Joseph, Neihardt's Black Elk, Brown's Black Elk, Luther Standing Bear, Lame Deer, Vine Deloria, and a number of other more-or-less recent Indian spokespersons are indiscriminately quoted as evidence for traditional American Indian conservation, ecology, environmental religion, or land ethics. Although I am not unmoved by these testimonials, a less sympathetic critic might dismiss them on several counts. They are for the most part relatively recent (mid-nineteenth century or later); they may be idiosyncratic (rather than a genuine reflection of shared cultural values); and they may be (quite understandably and justifiably, but nevertheless merely) nostalgic. As one is forcibly dispossessed of one's ancestral lands, it is natural to feel both outrage at colonial usurpers and nostalgia for one's rightful natural heritage.

The tender environmental sentiments in the testimonials one finds in such eclectic collections as T. C. McLuhan's *Touch the Earth* (1971) or Peter Nabokov's *Native American Testimony* (1978) may be only a personal reaction and cultural afterthought, a natural response to cultural oppression and personal dispossession.

A second approach to the question of a traditional American Indian land wisdom bases conclusions on what might be called descriptive ethnography. From a critical point of view, this method constitutes a quantum leap in sophistication and reliability compared with the testimonial technique.

The descriptive ethnographic method provides a firsthand account of the environmental behavior of contemporary remnants of hunter-gatherer cultures and a record of their accompanying beliefs, attitudes, and values as expressed both discursively and symbolically in ceremony, song, story, myth, and legend. Two outstanding examples of this method are Adrian Tanner's *Bringing Home Animals* (1979) and Richard Nelson's *Make Prayers to the Raven* (1983). Such accounts are culturally specific—often, as Tanner's work with the Mistassini Crees, subspecific—and critically disciplined.

The painstaking empirical research, dedication, and sympathy of contemporary ethnographers—whose work requires them to live in circumstances of hardship in necessarily remote and inhospitable environments—fills me, an armchair scholar, with unbounded respect and admiration. Still, the degree to which these contemporary accounts may be generalized not only across contemporary cultural boundaries, but across several centuries into the pre-Columbian past of the *same* culture is problematic. The influence of global technological civilization and the ideology that engendered it and in which it remains grounded is (except perhaps among a few Amazonian peoples) inescapable. It is also insidious. Technologies are never cognitively and axiologically neutral. They are embedded in an engendering and sustaining system of ideas. To buy guns, motors, and mackinaw jackets is to buy, however unintentionally, a worldview to boot. Both the Koyukons and the Mistassini Crees use guns and steel traps in taking game, wear some store-bought clothes, and to one degree or another are influenced by Christianity, modern science, Western medicine, money, and materialism.

On the other hand, what other assumption may be made than that those cultural elements not identifiably Western in origin must be at least a *legacy* of the aboriginal cultural past, if not identical with it? Still, an implacable skeptic might argue that the apparently reverential or ethical attitudes (whether express or implied) toward nonhuman natural entities are a consequence of Western influence no less than self-conscious conservation principles and practices like the family hunting territory. As I suggested earlier, conservation comprises a complex of ideas deeply intertwined with the foundational cognitive elements of the Western worldview. Reverential or ethical attitudes could also be interpreted as a dialectical reaction to typical Western attitudes of indifference and brutality toward nature. And as a bottom line, the interests of the ethnographers, their research agenda, the questions they pose to themselves or ask their informants may subtly shape the conclusions they draw. Contemporary ethnographers, it is pleasing to note, are very often also dedicated environmentalists. A critic might argue that research results documenting the existence of ethnic environmental attitudes and values congenial with the environmental attitudes and values of the investigator may be as much the product of a natural human bias and selectivity as of objective, disinterested, and balanced reporting.

To obviate at least some of the uncertainties of the descriptive ethnographical approach to the verification of the hypothesis that there existed some sort of environmental wisdom among traditional American Indians, Calvin Martin in *Keepers of the Game* (1978) employed a method he called ethnohistory. He suggested complementing ethnographic reports with historical documents—the letters, reports, chronicles, records, and so on, of explorers, traders, missionaries, European captives who lived among natives, acculturated natives, and such, as close to the documentary horizon as possible. Such documents portray Indian material and cognitive culture at first contact or soon thereafter, before generation upon generation of ever-increasing cultural influence from Europeans. The often casual and unsystematic and always ethnocentric and distorted quality of these early documents can then be compared and cross-checked with the more systematic and objective, but always relatively recent, ethnographic accounts in such a way that ideally they correct, supplement, enrich, and illuminate one another. Martin, in my opinion, makes very effective and persuasive use of this method to reconstruct a portrait of a "land ethic" type of land wisdom pervasive among eastern subarctic hunter-gatherers on the eve of European contact. (It is unfortunate that Martin discredited his own constructive profile of an aboriginal American Indian land ethic, however, with his eccentric and entirely speculative hypothesis of a war between Indians and animals and his petulant and self-contradictory effort to discredit the "myth" of an American Indian land wisdom perpetrated by contemporary environmentalists).

Finally, Thomas W. Overholt and I, in our book *Clothed-in-Fur and Other Tales* (1982), followed a method first proposed by A. Irving Hallowell (1955) similar to the methods of philosophical analysis and literary criticism. A putative traditional American Indian land wisdom in any interesting sense, from the point of view of contemporary environmentalism, would be not the personal wisdom of an exceptional Indian sage or philosopher, but the collective environmental ethos of a community. Such an ambient and possibly implicit aspect of a cultural worldview is borne, like all other aspects of cognitive culture, by an ambient and communal vehicle—a culture's language.

While language is routinely used day to day to convey personal meanings and messages, it is also the repository of a common system of cultural meanings and a common narrative heritage. Hence the systematic study of a culture's common medium (language per se) and narrative heritage (its general fund of myths, legends, and tales) should provide a reliable and objective method for a recovery of its common beliefs, attitudes, and values, including environmental beliefs, attitudes, and values. And though this necessarily remains a speculative matter, this method should give us a glimpse beyond the documentary horizon—to the extent that a culture's language survives and its narrative heritage lives on. Stories have, as it were, a life of their own. They persist with only incidental changes, through radically changed cultural circumstances. Consider the Euro-American oral heritage of fairy tales. They are about princes and princesses living in castles, knights in shining armor, magic swords, witches, sorcerers, and trolls. They hark back to another physical and psychic world. Yet they live on, relatively unchanged in the retelling, in our own world of skyscrapers, airplanes, computers, and technocrats.

Unlike Frank Speck, Irving Hallowell does not appear to have been especially interested in environmental issues or American Indian environmental attitudes and values. Hence his semantic analyses of the Ojibwa language cannot be impugned as consciously or unconsciously biased in favor of one or another type of American land wisdom. Overholt and I undertook a reex-amination of Hallowell's analysis of Ojibwa semantic categories with an eye to applying them to the question of an Ojibwa land wisdom. According to Hallowell (1960, 23–24; Black 1977), the formal Ojibwa linguistic distinction between animate and inanimate (analogous to gender distinctions in Romance languages) does not correspond to scientifically informed Western intuitions. For example, some stones (flint), certain kinds of shells (the megis shell of the midewiwin, for instance), thunder, various winds, and so on, as well as plants, animals, and human beings, fall into the animate linguistic class. Further, the category of "person," according to Hallowell, is not coextensive with the category "human being" in Ojibwa semantic discriminations as it is in English or other modern Western languages (Hallowell 1960, 21). Animals, plants, stones, thunder, water, hills, and so on may be "persons" in the Ojibwa linguistic organization of experience.

Now, as Hallowell points out, there is an intimate link in Ojibwa, as in English, between persons and a complex network of social interaction. Since nature is more broadly animate and personal in the traditional Ojibwa world than in the contemporary Western world, "the world of personal relations," according to Hallowell (1960, 43), "in which the Ojibwa live is a world in which vital social relations transcend those which are maintained with human beings." The Ojibwa cycle of myths details, elaborates, and amplifies the personal and social organization of nature and of human/nature interaction structurally represented in Ojibwa semantic discriminations. Using William Jones's extensive and remarkable collection of stories, *Ojibwa Texts* (1917–19), Overholt and I found that the Ojibwa narratives consistently represent the natural world as a world of other-than-human persons organized into a congeries of societies. Plant and animal species are, as it were, other tribes or nations. Human economic intercourse with other species is represented not as the exploitation of impersonal, material natural resources, but as reciprocal gift giving or bartering, in which both the human and nonhuman parties to the exchange benefit. Game animals give their skins and flesh to human beings, who in return give the animals tobacco and other desirable cultivars and artifacts. The slain animals are reincarnated in the most literal sense of that term—reclothed in flesh and fur—and thus come back to life to enjoy their humanly bestowed benefits.

The *nomoi*, the rules or customs governing human/nature relationships among the Ojibwas, are thus of an essentially social-ethical sort. The animal spirits are not worshiped in any religious sense of the term; rather, like members in good standing of human society, they are respected. Their personal interests and feelings are taken into account. As in any social ethic, the rules or conventions serve to formalize and articulate this affective moral posture of respect for persons and to provide behavioral guidance.

Most interesting, we found a correspondence in abstract form between the Ojibwa environmental ethic and the Aldo Leopold land ethic, which is the most celebrated and appealing version of environmental ethics in contemporary environmental thought (Callicott 1982). The principal idea upon which Leopold rests his land ethic is the ecological concept of a biotic community: "All ethics so far evolved rest upon a single premise: that the individual is a member of a community of interdependent parts. . . . The land ethic simply enlarges the boundaries of the community to include soils, waters, plants, and animals, or collectively, the land" (Leopold 1949, 219).

The detailed representations of the personal-social order of nature among the Ojibwas, on the one hand, and among contemporary ecologists like Aldo Leopold, on the other, are of course vastly different. The one is mythic and anthropomorphic, while the other is scientific and self-consciously analogical. Nevertheless, when the mythic and scientific detail is stripped away from either, an identical abstract structure—an essentially social structure—constitutes the core conceptual pattern of the totemic natural community of the Ojibwas and the biologist's economy of nature. In form, thus, the Ojibwa land ethic and the Aldo Leopold land ethic are identical.

On the basis of these reflections, what conclusions may be reached? Did traditional American Indians possess an environmental wisdom? If so, of what type? How can we be sure whether they did or not at this distance from a past we cannot autopsy?

To take up the last question first, contemporary descriptive ethnography, ethnohistory, and ethnolinguistic/narrative analysis are all useful methods and may be cooperatively brought to bear on the question of a traditional American Indian land wisdom. For example, in the tradition I am most familiar with, the Algonkian ethnographies of Speck and more recently Tanner, the eastern subarctic ethnohistory of Martin, the more analytic ethnometaphysical studies of the Ojibwas by Hallowell, and the Ojibwa texts of

Jones all agree in fundamental particulars about the cognitive organization of the Ojibwas and more generally the Algonkian world. Nonhuman natural entities are personal beings, socially organized into families, clans, and nations not unlike the traditional Algonkians themselves. Relations with these other-than-human persons accordingly are socially structured. They are courteous, cautious, mutual, reciprocal, deferential, diplomatic—forms of conduct that must be maintained to sustain the interspecies social structure and, so to speak, international balance of power. From a sociobiological point of view, this is the sum and substance of an ethic—an American Indian land or environmental ethic.

To generalize a priori beyond the Ojibwa and Algonkian cultural materials I have mentioned would not be warranted. A similar collection and comparison of ethnography, early historical documents, and semantic and mythic analyses from other American Indian cultures would almost certainly reveal very different cognitive organizations of phenomenal experience and correspondingly and proportionately different attitudes and values respecting nature. What was the traditional Papago, Shoshone, Hopi, or Cherokee land wisdom, if any, like? If a concerted and systematic effort were undertaken, an ethnographic map of types of American Indian land wisdom might gradually be built up and, moreover, correlated with the biogeographical map of North America, in the way Nelson (1982) has suggested, to develop a kind of North American ecology of mind.

To speculate briefly on other Plains cultures, if the Lakota worldview familiar to everyone from *Black Elk Speaks* withstands critical scrutiny, then the Sioux pictured nature as more like a vast extended family than a congeries of societies. Such a worldview appears to be corroborated by the Lakota mythic materials collected in the 1890s by Walker (1980). An environmental wisdom is certainly immediately inferable from such a representation, but it would not be very precisely described as an ethic. One's familial duties, it seems to me, go beyond ethics. Ethics suggests, at least to me, a formality inappropriate to intimate family relations. No matter how such tender environmental sentiments as those Black Elk expresses may be labeled, however, there seems to me to be an important environmental consciousness and conscience implicit in them.

As a philosopher, professionally concerned not only with conceptual analysis and clarity but also with consistency, I am inclined to think that conceptually inconsistent environmental attitudes cannot be credibly attributed to a single people. In particular I would suppose, as I remarked earlier, that prudential utilitarian conservation, since it is conceptually dissonant with either a religious or an ethical intellectual orientation to nature, could not be autochthonous to a culture in which a religious or ethical orientation to nature was manifest. Reflection stimulated by correspondence with Richard Nelson, however, has convinced me that this judgment is wrong. In a letter to me (May 2, 1985), Nelson wrote:

> Regarding empirical knowledge of ecological processes, sustained yield practices based on such knowledge, and deliberate conservation of resources . . . [m]y experiences with Koyukon people especially (and Kutchin to a lesser degree), convince me that they had a well-developed, empirically based system of ecological knowledge and conservation prior to contact with Europeans. It is definitely my impression that people like the Koyukon approach their environment as much from a scientific, empirical point of view, as from an animistic and symbolic perspective. This includes not only the practical, technical processes of locating and obtaining food, but also the longer range processes of maintaining food resources. For the Koyukon and Kutchin, conservation of these resources is definitely a conscious goal.

Nelson here says in effect that the traditional Koyukons and Kutchins were ecologically informed, deliberate conservationists, as well as animistic environmental ethicists. But how could such disparate beliefs coexist in one worldview?

A little reflection on the Western worldview to which I was enculturated suggested that such conceptual incongruities may be, despite philosophers' apparently aberrant passion for consistency, the human norm. Christianity coexists more or less comfortably with modern science. Euro-Americans function in an empirical, technical, material, and mechanical world, and at the same time many believe in God, heaven, hell, miracles, possession, the power of prayer, life after death, and the other general doctrines of Christianity.

Why should I be reluctant to suppose, therefore, that the traditional Koyukons or Ojibwas could at once represent their environments as animate, personal, and social *and* as a systemically integrated pool of impersonal resources that must be calculatively sustained and prudently conserved? Typical modern Euro-Americans can believe that a rich man has as little chance to enter heaven as a camel has to pass through the eye of a needle, yet spend their lives in pursuit of the Almighty Dollar. At least among the Koyukons and Ojibwas the behavioral implications of conservation and land ethics converge, however dissonant their cognitive foundations. Both serve to inhibit thoughtless destruction and overexploitation of nature. They are behaviorally complementary, if not conceptually consistent. On the other hand, Western religious beliefs and Western scientific-technological materialism are often contradictory in their practical implications as well as mutually inconsistent in their cognitive foundations. But that doesn't stop many Euro-Americans from espousing both.

Of the several kinds of land wisdom attributed to traditional American Indian cultures, I am most skeptical of the religious reverence type, at least as applied to the Ojibwa and similar Indian worldviews. I wish to emphasize immediately, however, that my concerns are more terminological than substantive. To say that American Indians had a spiritual representation of nature is misleading because of the *usual connotation* of the term "spiritual," which suggests an otherworldly and eschatological orientation of mind. To believe that animals, plants, wind, thunder, rocks, mountains, the sun and moon, and so on are conscious and communicative persons, like ourselves, and to hypostatize or reify such consciousnesses, including our own, as owing to the presence of souls, spirits, or manitous is not quite the same as a "spiritual" orientation to life in the *ordinary Western sense of the term*. Western religious thought has been overwhelmingly dualistic. The spirit and flesh not only are ontologically distinct, they are morally opposed from the point of view of traditional Western religious thought. Hence the term "spiritual" ordinarily implies an antinatural moral opposition to the "carnal," "brute," physical world—as it is regularly stigmatized from a conventional "spiritual" point of view. It is equally misleading, I think, to characterize the means of communicating with nature spirits—dreams, visions, divination, and ceremonials—as "religious rites," given the usual connotation of the term "religious." Hence, for all its rhetorical power and stylistic appeal, Speck's declaration that hunting was a "holy occupation" among Algonkian hunter-gatherers misses the mark. From the data he provides, it seems to have been, rather, a magical and moral occupation. Human life in nature from the perspective of the Lakota organization of experience as portrayed in *Black Elk Speaks* and *The Sacred Pipe*, on the other hand, might more accurately be characterized as religious or holy, since prayers and worshipful rites seem to figure more prominently in the Lakota ideal of human-nature relationships than in the Algonkian.

Finally, we may ask, can a traditional American Indian land wisdom help to guide the United States and other modern nations out of the present environmental malaise? I think certainly it can. If Richard Nelson is correct to suppose that some traditional American Indian peoples practiced conservation complemented by a land ethic and maintained a long-term balance between themselves and nature, then in his words, "If they can do it, so can we" (letter to me, May 2, 1985). Their example, I think we may be confident, represents hope. It also represents a role model, despite Calvin Martin's disclaimer, that we can relate to and emulate. If Overholt and I are right to suppose that some American Indian peoples portrayed their relationship with nature as essentially social and thus, by implication, as essentially moral, then their rich heritage could provide, ready-made, the myths and the parables missing from abstract articulations of a biosocial environmental ethic like Aldo Leopold's.

A traditional American Indian land or environmental wisdom is not a neoromantic invention. But we are just beginning to explore what it actually amounted to, and only in some cultures and in some bioregions. More careful, systematic, and critical research will almost certainly prove enormously rewarding and, ultimately, of perhaps the greatest possible practical benefit to contemporary society.

REFERENCES

Bishop, Charles A. 1970. The emergence of hunting territories among the northern Ojibwa. *Ethnology* 9:1–15.

Black, Mary B. 1977. Ojibwa taxonomy and percept ambiguity. *Ethos* 5:90–118.

Callicott, J. Baird. 1982. Traditional American Indian and Western attitudes toward nature: An overview. *Environmental Ethics* 4:293–318.

Grieder, Terrence. 1970. Ecology before Columbus. *Americas* 22:21–28.

Guthrie, Daniel. 1971. Ojibwa ontology behavior and world view. *BioScience* 21:721–23.

Hallowell, A. Irving. 1955. *Culture and experience.* Philadelphia: University of Pennsylvania Press.

———. 1960. *Ojibwa ontology, behavior, and world view.* In *Culture in history: Essays in honor of Paul Radin*, ed. Stanley Diamond. New York: Columbia University Press.

Hughes, J. Donald. 1977. Forest Indians: The holy occupation. *Environmental Review* 2:2–13.

Jones, William. 1917–19. *Ojibwa texts*. 2 vols. New York: Publications of the American Ethnological Society.

Kaiser, Rudolph. 1987. A fifth gospel, almost: Chief Seattle's speech(es)—American origins and European reception. In *Indians and Europe: An Interdisciplinary Collection of Essays*, ed. Christian F. Feest. Aachen: Rader Verlag.

Leacock, Eleanor. 1954. *The Montagnais "hunting territory" and the fur trade*. Memoir 78. Washington, D.C.: American Anthropological Association.

Leopold, Aldo. 1949. *A Sand County almanac, and sketches here and there*. Oxford: Oxford University Press.

McCleod, William Christie. 1936. Conservation among primitive hunting peoples. *Scientific Monthly* 43:562–66.

McLuhan, T. C., ed. 1971. *Touch the earth: A self-portrait of Indian existence*. New York: Outerbridge and Lazard.

Martin, Calvin. 1978. *Keepers of the game: Indian-animal relationships and the fur trade*. Berkeley: University of California Press.

Nabokov, Peter, ed. 1978. *Native American testimony*. New York: Harper and Row.

Nelson, Richard K. 1982. A conservation ethic and environment: The Koyukon of Alaska. In *Resource managers: North American and Australian hunter-gatherers*, ed. Nancy Williams and Eugene Hunn. Boulder, Colo.: Westview Press.

———. 1983. *Make prayers to the Raven: A Koyukon view of the northern forest*. Chicago: University of Chicago Press.

Overholt, Thomas W. 1979. American Indians as "natural ecologists." *American Indian Journal* 5(9): 9–16.

Overholt, Thomas W., and J. Baird Callicott. 1982. *Clothed-in-Fur and other tales: An introduction to an Ojibwa world view*. Washington, D.C.: University Press of America.

Reichel-Dolmatoff, G. 1976. Cosmology as ecological analysis: A view from the rain forest. *Man* 2:307–18.

Richie, William A. 1955–56. The Indian and his environment. *New York State Conservationist* 10:23–27.

Speck, Frank G. 1938. Aboriginal conservators. *Bird-Lore* 4:258–61.

———. 1939. Savage savers. *Frontiers* 4:23–27.

Starnes, Luke. 1968. The sage of Seattle. *Golden West* 4:34–37, 60–64.

Tanner, Adrian. 1979. *Bringing home animals: Religious ideology and mode of production of the Mistassini Cree hunters*. New York: St. Martin's Press.

Udall, Stewart. 1972. First Americans, first ecologists. In *Look to the mountain top*, ed. Charles Jones. San Jose, Calif.: H. M. Gousho.

Walker, James R. 1980. *Lakota belief and ritual*. Ed. Raymond DeMallie and Elaine Jahner. Lincoln: University of Nebraska Press.

Chapter Eleven

Plains Indian Influences on the American Environmental Movement: Ernest Thompson Seton and Ohiyesa

O. Douglas Schwarz

A high regard for the Native American is a recurrent theme in the contemporary American environmental movement. Indians are perceived as having lived in harmony with nature before the European invasion of the continent. Many environmentalists believe there are important ecological lessons to be learned from studying the traditional life-styles of the native peoples. In addition, the philosophies/religions of the American Indians are often looked upon as models of thought that the industrialized West might do well to emulate. Native American worldviews have been compared favorably with Aldo Leopold's "land ethic" (Callicott 1982 and this volume), perhaps the highest compliment possible from the environmentalist perspective.

In recent years, however, the validity of this image of the American Indian has been questioned. The scholarly work cited most frequently in opposition to the ideal of the "ecological Indian" is Calvin Martin's *Keepers of the Game* (1978), which contends that the traditionally respectful attitude of Native Americans toward nature was a matter of self-interest rather than some sort of environmental ethic. As Baird Callicott has pointed out in the preceding chapter, much of Martin's material actually seems to support the notion that the traditional Eastern Algonkian culture he studied contained a type of environmental ethic. I can only agree with Callicott that it is unfortunate Martin chose to interpret the Algonkians' respectful behavior toward game animals as a manifestation of fear and hostility rather than of genuine regard for the animals as fellow beings, and Martin has been criticized for this elsewhere as well (see Krech 1981).

Perhaps more important, however, Martin argues that even if some natives *were* guided by an environmental ethic, that ethic could not possibly be adopted by the Western environmental movement, since its principles would be utterly alien to those of Western thought (Martin 1978, 188n). Though Martin was not writing primarily of the Plains tribes, his arguments were intended to apply to them.

How is it, then, that environmentalists tend to hold American Indians in such high esteem? Martin seems to believe this attitude is recent. Time and again he refers to the emergence of the Native American as the "spiritual leader of the ecology movement" as an event that took place in the late 1960s: "The idea would never have taken hold had it not been that conservationists needed a spiritual leader at that particular point in time, and the Indian, given the contemporary fervor and theology of environmentalism, seemed the logical choice" (Martin 1978, 157; see also p. 19 and Martin 1981, 137). He insists, however, that those who "canonized" the Indian during the sixties had no real understanding of native attitudes toward nature. If they had, they would presumably have realized, with Martin (1981), that "to suggest that we [Westerners] might adopt . . . an Indian world view is absurd." As I mentioned above, Martin is not without his critics, but most of the criticism has focused on his views as to the content of native attitudes toward nature. Scant attention has been paid to his claim that those attitudes, whatever they might be, could never have meaning for western environmentalists. Thus this critical aspect of Martin's thesis remains largely unchallenged and has even gained some acceptance in environmental circles (Regan 1982; Steinhart 1984, 9).

His position that modern industrial humanity cannot learn ecological lessons from Indians is scarcely immune to criticism, however. The fact is that we have already learned from tribal societies in this area. It is easy to show that respect for native traditions is a long-established characteristic of the American environmental movement, not a phenomenon born of the turbulence and "fervor" of the 1960s. It can further be shown that native traditions—particularly those of the Plains peoples—directly influenced certain leaders of the movement, contributing to the development of their own environmental philosophies. Moreover, at least some of those who were so influenced were well acquainted with native cultures, so they cannot be dismissed as victims of some romantic image of Indians as ecologically noble savages.

In the preceding chapter, Baird Callicott performed the valuable service of defining what sorts of meanings may attach to various (generally non-Indian) notions of Indian "land wisdom." I shall be less concerned with the precise content of native "ecological" values than with the simple fact that such values have been recognized, admired, and adopted by various Western environmentalists—despite Martin's insistence that this could never happen.

A list of leading environmentalists whose lives and ideas were influenced, to varying degrees, by contact with Native Americans and their worldviews reaches back to the earliest days of the movement (the beginnings of such a list appear in Fox 1981, 350 ff.). The historian Robert Sayre, for example, documents a significant native influence on the work of Thoreau, who hoped to write a book about Indians had he lived long enough. According to Sayre, "Thoreau recognized Indians as people who had spent their lives in Nature and developed a knowledge of it that was superior to white man's. . . . Though we can say that he was more involved with the idea of 'the Indian' . . . than with Indians as people, we cannot dispute the extent of the involvement" (Sayre 1977, x, 203).

As Sayre indicates, Thoreau may well fall into the category of environmentalists who have admired Indians from a position of relative ignorance. Another who might fit this description is John Muir, who "occasionally brushed against Indian and Eskimo cultures, and sensed a corresponding affinity with their religious ideas" (Fox 1981, 80). Not all of their compatriots, however, can be so easily dismissed.

John Wesley Powell gained fame through his explorations of the Grand Canyon, but for conservationists his 1878 "Report on the Lands of the Arid Regions of the West" stands as the first comprehensive plan for the ecologically sound management of the western territories. Powell was also an avid student of native cultures. He "knew more of the live Indian than any other man" (Udall 1963, 88) and "accepted the right of the Indians to their own customs and beliefs" (Strong 1971, 45).

Mary Hunter Austin, author and activist for the cause of conservation, was directly inspired by the life-styles of the Hopis, Paiutes, and other native peoples of the Southwest. She adopted their mode of dress and sought to emulate their spiritual life as well (Wild 1979).

Robert Marshall, associated with both the United States Forest Service and the Wilderness Society, also ran the United States Office of Indian Affairs for many years. He wrote that Indians "had for the most part an attitude of preservation, a realization that their life came from a nature which it would be catastrophic to destroy" (Fox 1981, 350).

Gifford Pinchot, who coined the term "conservation," considered the hunting methods of the Algonkian peoples to represent "conservation practice at its best" (Fox 1981, 350). Ernest Oberholtzer of the Wilderness Society "immersed himself in Indian cultures, adopting their standards of beauty and tape-recording Ojibway religious myths" (Fox 1981, 364).

Finally, Aldo Leopold, developer of the "land ethic" that forms the philosophical basis for much of modern preservationist thought, was also inspired in part by a Native American example. Early in his career he observed that the Apache peoples of the White Mountain region of Arizona and New Mexico had lived in that fragile ecosystem for centuries without damaging the environment. This contrasted sharply with the ecologically ignorant practices of the newly arrived whites, who were rapidly destroying the land (Wild 1979, 95).

This litany could easily be continued into the latter half of the twentieth century, embracing such modern environmentalists as Stewart Udall, William O. Douglas, Sigurd Olson, and Peter Mathiessen.

Several early environmentalists were specifically influenced by their acquaintance with the Plains tribes. We know George Catlin, for example, primarily for his 1830s studies of the "manners, customs, and conditions" of various Indian peoples, particularly the Plains tribes, but he was also an early advocate of environmental protection. Catlin foresaw the extinction of the buffalo and, with them, the passing of the peoples of the Great Plains, and even of the Plains themselves. In calling for a national park that would preserve the Plains, the buffalo, and the Indians—the world's "noblest specimens" of beast and man in the world's most "beautiful and lovely scenes"— Catlin was the first to suggest that the federal government take a hand in preserving America's unique biotic and cultural heritage. His efforts contributed to the establishment of the first national park (though outside the Plains) at Yellowstone (Nash 1976, 5–9).

Francis Parkman, author of *The Oregon Trail* and many other volumes of history concerning the American West, lived for a time with the Oglala Sioux. Stewart Udall writes: "This Oglala interlude, at twenty-three, was the summit of Parkman's active life. The simple ways of the Sioux and their attachment to the land reverberated in his mind as long as he lived" (Udall 1963, 45–46). Parkman's Oglala experience contributed to his becoming a proponent of both environmental protection and Indian rights (Doughty 1962, 316). Like Catlin, he recognized that the fate of the American land— particularly the Plains—was intimately linked with the fate of its native peoples. He perceived that "great changes are at hand in that region [the Plains]. . . . the buffalo will dwindle away, and the large wandering communities who depend on them for support must be broken and scattered" (Parkman 1964, 163–64). Indeed, he consciously sought to chronicle "the American forest and the American Indian at the period when both received their final doom" (Doughty 1962, 94).

A more important figure in the environmental movement who was influenced by Plains traditions was George Bird Grinnell, on whom I was tempted to focus this chapter. Grinnell championed the cause of conservation as editor of *Forest and Stream* magazine from 1881 to 1911, was instrumental in establishing Yellowstone and Glacier national parks, helped found both the New York Zoological Society and the first Audubon Society, and was also among the foremost experts of his day on American Indians. He had lived in a number of Plains societies, most notably the Cheyenne and Pawnee, becoming a member of the latter tribe. In later years he wrote studies of the Plains peoples that were definitive in their time and are still classics of Native American ethnography. Grinnell clearly admired certain aspects of the Indians' attitude toward nature, for example native hunting practices and their deliberate conservation of game animals. These he thought worthy of emulation by white sportsmen (Fox 1981, 350).

There is, however, an even more interesting example than Grinnell. Ernest Thompson Seton, to be sure, has fewer credentials than Grinnell as either an environmentalist or a Native American specialist. On the other hand, his case is particularly dramatic in that he not only drew much of his own life's inspiration from native (and primarily Plains) sources, but actively sought to promote Indian ways among others. In this effort, furthermore, he had the guidance and assistance of a distinguished member of a

Plains culture, Dr. Charles Alexander Eastman (Ohiyesa) of the Santee Sioux. Ernest Thompson Seton contributed to the American environmental movement in many ways. He was a naturalist and wildlife illustrator who made important contributions to North American zoology, he helped found the Boy Scouts of America—an organization that has introduced generations of young men to the outdoor life—and though a citizen of Canada for much of his life, he served as Theodore Roosevelt's informal ambassador to that country on environmental matters (Seton 1967, 132–34). He lectured widely on conservation issues, arguing on both moral and practical grounds for the importance of preserving wilderness and species diversity (Seton 1967, 165–68). With all of this to his credit, however, he was probably best known as an author of popular animal stories consciously designed to teach respect for the rights of animals (Seton 1898, 9–10).

Seton was also all but obsessed with the study of Native Americans, whom he considered the most "heroic" and "physically perfect" race of humanity. He credited them with creating "the most spiritual Civilization the world has ever seen" (Seton and Seton 1966, 107–8). A native of Scotland, he seems to have derived his early love of Indians from literature, almost certainly including the works of James Fenimore Cooper, and by the age of fourteen his dream was "to emulate the Redman." He tried to organize his boyhood gang into an Indian tribe (but found them more taken with Robin Hood). Failing that, he built an "Indian cabin" in the woods and contrived to spend as much time as possible living life in the wild, an experience he believed gave him much wisdom (Seton 1940, 100–109).

Obviously, Seton as yet knew little of Indians—but by his early twenties he had come to North America and begun to experience firsthand what he had only read and dreamed about before. He traveled extensively in the West, encountering the Assiniboines, Sioux, and other Plains peoples, and he developed close friendships with any number of natives, most notably a Cree hunter named Chaska and the Sioux physician and author Charles Alexander Eastman.

Eastman's story is probably familiar to most scholars of the Plains and its peoples. Born on the Santee reservation in 1858, he was separated from his family during the Minnesota Sioux uprising of 1862, in which his father was presumed killed. The four-year-old Ohiyesa was taken to Canada by relatives, where he lived in the traditional manner of his people until he was fifteen. His father then reappeared—now a homesteader in the Dakota Territory—to reclaim his son and teach him the ways of civilization. Ohiyesa's success in adapting to those ways was remarkable, leading him to Dartmouth College in New Hampshire and ultimately to a medical degree. His marginally successful medical career, however, was seldom as important to him as his work to improve the lot of the Indian peoples and to preserve and transmit native ideals, and he may be the first successful Native American author, with many books and articles to his credit, though the editorial efforts of his wife, Elaine Goodale, were probably essential to his success (Miller 1978, 66). He also lectured widely.

For Eastman, part of what native traditions had to offer to the conquering white civilization was their greater awareness of and respect for the environment, embodied in the teachings of the native religions.

> We believed that the spirit pervades all creation and that every creature possesses a soul in some degree, though not necessarily a soul conscious of itself. The tree, the waterfall, the grizzly bear, each is an embodied Force, and as such an object of reverence. . . . [The Indian] recognizes the spirit in all creation, and believes that he draws from it spiritual power. His respect for the immortal part of the animal, his brother, often leads him so far as to lay out the body of his game in state and decorate the head with symbolic paint or feathers. (Eastman 1980, 14–15, 47)

Given such an attitude, it was natural for Eastman to become involved in the "back-to-nature" movement of the early 1900s—and thus to become a cohort of Ernest Thompson Seton.

Perhaps partly due to Eastman's influence, Seton became enamored of the religions of the Native American peoples—or rather, of the fundamental religious outlook he believed most tribes shared. Seton (1954) himself saw visions, heard voices, and claimed to be guided by a force he called the Buffalo Wind, which he associated with the traditions of the Native American peoples (see also his autobiography [Seton 1967, 146–47]). At age forty-five he received a "message" through an American woman who had studied under "the Great Masters" in India: "You are not of this people [the white race]. Why do you not open your ears and your heart? Know you not you are a Red Soul sent back to deliver The Message—to show forth the redman way? Hear it now—for the Buffalo Wind is blowing!" (Seton 1954,

396).[1] According to Julia M. Seton, his second wife, Seton never ceased to concentrate on the "mission" he had been given—and Seton himself writes, "My lifelong dream and hope is that I may be the instrument of giving to the whiteman's world the inspiring teachings of the Redman" (Seton and Seton 1966, 104).

Seton's admiration for Native Americans is most readily apparent in one of his last books, *The Gospel of the Redman*, which frankly proclaims the superiority of native over white culture (p. 108). Though primarily concerned with Indian religion, the book also devotes attention to the conservationist tendencies of the native peoples. Among the "fundamental laws" of the Indians he thought worthy of emulation by whites were precepts against killing animals for pleasure, against exterminating species, and against wasting any part of the game one killed (Seton and Seton 1966, 30). Clearly, respect for nature and conservationist tendencies were at least part of what made the Native American, in Seton's view, "the apostle of outdoor life, [whose] example and precept are what the world needs to-day above any other ethical teaching of which I have knowledge" (Seton and Seton 1966, xvi).

Seton's most ambitious attempt to convert America to the "gospel of the redman" was his establishment of a boys' organization (later expanded to include girls) known as the "Seton's Indians," the "Woodcraft Indians," and finally the "Woodcraft League." Since he believed his own boyhood experience of living "like an Indian" had benefited him, Seton now proposed to mold other young men in the image of his Indian ideal. To do this he borrowed liberally from the traditions of the native American peoples, filling his *Book of Woodcraft* with native stories, songs, games, and lore—supplemented by his own outdoor experience and zoological knowledge. The sources of his Indian lore, when acknowledged (both here and in *The Gospel of the Redman*), demonstrate the extent of his personal contact with natives as well as his familiarity with both popular and ethnographic literature. He quotes reports of the Bureau of American Ethnography; such notables as photographer Edward S. Curtis, anthropologist Alice C. Fletcher, George Catlin, and George Bird Grinnell, along with lesser white authorities; and native authors such as Long Lance and Standing Bear—and he spices it all with personal anecdote.

Charles Eastman is given general credit as a source for the *Book of Woodcraft*, and he is specifically thanked for offering general criticism and "special assistance" with three chapters of the book (Seton 1923, vii).[2] Eastman also served on the board of the Woodcraft League (Dee Seton Barber, letter to me).

Seton's league provided part of the inspiration for the Boy Scout movement established in England by Lord Baden-Powell (Levy 1944, 18). In due course, the league was absorbed into the Boy Scouts of America, though Seton eventually became disillusioned with the new organization and withdrew (Levy 1944, 26). Seton's influence is still evident, however, and the Boy Scouts of America still distributes some of Seton's works, including *The Gospel of the Redman*.

The extent to which scouting has contributed to environmental awareness in America would be difficult to quantify, though that there has been some contribution seems self-evident. The influence of Seton's league, however, does not end with the Boy Scouts. The founders of the Campfire Girls and the YMCA, for example, both sat on the board of the Woodcraft League in earlier years (Dee Seton Barber, letter to me),[3] and both organizations show the league's influence. As with the Boy Scouts, one must assume that the Campfire Girls and the YMCA (with its summer camp programs) have made at least some contribution to environmental awareness in America.

To some extent the continuing influence of Seton's league can be attributed to Charles Eastman, who worked with the Scouts, the Campfire Girls, and the YMCA at various times after the demise of the league (Miller 1978, 62, 66). But Eastman not only worked to spread the "gospel of the redman" within existing organizations, he also struck out on his own by establishing girls' and boys' summer camps on Granite Lake in New Hampshire, where the teaching of Indian lore was naturally a specialty. The "aims and ideals" of these camps, according to a brochure, were "in line with the best thought of today, which is strikingly in harmony with the original philosophy of the Native American." There can be little doubt that Seton's thought was among "the best of the day," in Eastman's estimation. The programs outlined in camp brochures, including weekly council fires and the like, are strongly reminiscent of the practices of the Woodcraft League. Seton, for his part, publicly endorsed the camps.

Eastman's camps operated for less than a decade, but the Woodcraft League practices he promoted were apparently adopted by other summer camps in the area. (Doubtless there may be other routes by which the Woodcraft League teachings became part of the common lore of the New

England summer camp circuit, but Eastman's camps surely contributed to this process.) I myself attended a New Hampshire summer camp as a boy (from 1963 to 1966), where many of the activities were derived from Seton's *Book of Woodcraft*. The Omaha tribal prayer we sang at our weekly "council fires," for example, was of immense importance to Seton, appearing first among the "woodland songs, dances, and ceremonies" contained in the *Book of Woodcraft* (Seton 1923, 61). It is also found in *The Gospel of the Redman*, with the note that those of "the Woodcraft Way" use the prayer just as the Omahas once did (Seton and Seton 1966, 21). Seton himself invariably used this prayer to close his own "grand councils" (Dee Seton Barber, letter to me), just as it was used at my summer camp.

Without question, my summer camp experiences had a profound effect upon my life and represent the earliest roots of my interest in both environmental issues and Native American traditions. In a very real sense I am an environmentalist (and an admirer of Indians) because of Ernest Thompson Seton and Ohiyesa's work. This, I suggest, is one concrete example of how the Plains Indian tradition has influenced and continues to influence the American environmental movement.

A critic might comment at this point that Seton was a hopeless romantic. Granting that he had some legitimate knowledge of native traditions and peoples, a man who contends that the American Indians represent humanity's nearest approach to perfection is obviously looking at his data through more than slightly rose-tinted glasses. Even Eastman, for all he was born and raised in the Sioux tradition, can be accused of romanticizing his own culture to some degree (Miller 1978, 63–64). Thus a Calvin Martin might argue that Seton was indeed ignorant—almost willfully so—of the realities of Indian tradition, despite his considerable exposure to natives and their ways. In Seton's defense, however, it may be said that he was at least conscious of his romanticism. He was well aware that not all Indians lived up to his ideal—but he believed that some did, and he deliberately chose to ignore the rest. In both the *Book of Woodcraft* (Seton 1923, 10) and *The Gospel of the Redman*, for example, he clearly states his intention to present only "the highest and best" aspects of native tradition. "My watchword, is 'The best things of the best Indians' just as we ourselves hope to be represented by our best brains and kindest lives, not by the ruffians and outlaws that form so large a part of our population" (Seton and Seton 1966, xv–xvi). These works are admittedly polemical; Seton sought sympathy, and more, for the "Redman Way." But an author with a strong point of view, and even a "romantic" approach to writing, may nonetheless have some knowledge of what he is talking about.

Calvin Martin believes that the environmental movement turned to the American Indian for spiritual leadership during the social turmoil of the 1960s and contends that those environmentalists who look to native traditions for guidance and inspiration do so out of ignorance. However, throughout the history of the American environmental movement there have been those who admired the native peoples and their traditions. The brief survey of such individuals offered here gives only a hint of how their environmentalism may have been shaped by their knowledge of and contact with native cultures. I hope, however, that this is sufficient to demonstrate that there *were* indeed such people long before the 1960s, and that not all of them were ignorant of the facts about native attitudes toward nature. Certainly George Catlin and George Bird Grinnell cannot be dismissed as know-nothings.

Ernest Thompson Seton went further than most in seeking inspiration from Native American traditions. Seton not only admired natives, he sought to emulate them in significant ways and worked to spread certain elements of native tradition—including attitudes toward nature—among whites. Like Catlin and Grinnell, Seton was hardly ignorant of the realities of the traditions he admired. Indeed, his views were shared by a prominent member of the Plains tradition, Charles Eastman, who was both a source of inspiration for Seton and a proponent of his ideas.

Calvin Martin might well disagree with Seton's interpretations of native tradition—as for that matter he might disagree with George Catlin, George Bird Grinnell, and even Charles Eastman. But he could not, I think, reasonably contend that none of these men (or any of the other environmentalists who have admired Native Americans over the decades) knew enough about native traditions to be entitled to an opinion about them.

Martin is wrong, then, in contending that knowledgeable Western environmentalists cannot possibly adopt Native American views of nature. It is instructive to ask how he reached this conclusion and just where his reasoning may have gone awry. As I stated earlier, Martin believes that native and Western views of nature are hopelessly incompatible. In his opinion this incompatibility is a function of the fact that both are ultimately rooted in the religious worldviews of the respective cultures. Native worldviews, according to his analysis, generally combine various aspects of what Western theology would call animism, pantheism, and polytheism—in sharp contrast to the monotheism of Western religion. Thus the fundamental reason native views of nature must forever remain inaccessible to Westerners is that those views constitute "an alien ideology . . . which is antithetical to the central dogmas of Christianity. And [Martin] simply cannot imagine how the two views are capable of reconciliation" (Martin 1978, 188 n.e).

Martin is quite correct, I think, in focusing on the essentially religious nature of traditional native attitudes toward the environment. Here I must part company with Baird Callicott, who (in the preceding chapter) hesitates to use terms such as "religious" and "spiritual" to describe the attitudes he has studied. Yet our differences on this point appear to be largely semantic. Certainly Callicott acknowledges that the Ojibwas' attitudes are at least in part a function of their belief in the presence of "souls, spirits, or manitous" in many of the creatures, structures, and forces of nature, and that "dreams, visions, divination, and ceremonials" are central to the Ojibwas' relationship with these presences. Few scholars of religion, I think, would hesitate to characterize such a worldview as religious. Callicott's concern that the term "religion" implies *dualistic, Western-style* religion seems to me unfounded.

To be sure, there can be little doubt that the religious worldviews of the various native peoples are quite different from the traditional Christian view that dominates Western society. It does not necessarily follow, however, that environmentalists from a Western background cannot accept at least some of the teachings of Indian traditions. If nothing else, psychologists building upon the theories of cognitive dissonance first proposed by Festinger (1957) have shown that it is not at all uncommon for people to believe in the truth of propositions (especially religious propositions) that seem inconsistent with their cultures' and even their own personal worldviews (Malony 1976).

Yet there may be less radical approaches to understanding how some Westerners—especially environmentalists—might be able to embrace Native American views of nature. For one thing, many of the environmentalists who have admired native traditions over the decades have not been

Christians, at least in Martin's sense. Martin himself recognizes Emerson, Thoreau, and Muir as "dissenters" whose spiritual paths diverged from the mainstream of Western religion (Martin 1981, 139). Of course these men do not represent the mainstream of Western religious thought. But it may be that their views do represent a significant countercurrent of religious thought that is not uncommon among environmentalists. Historian Stephen Fox has argued that a kind of covert "paganism" has been and continues to be shared by many environmentalists. "All but invisibly, often apologetically, ever since Thoreau naturalists and conservationists had embraced a variety of non-Christian religions. Sometimes the faith had no specific name but merely embodied (as Ansel Adams described his attitude) 'a vast impersonal pantheism' " (Fox 1981, 363). If Fox is correct, the "dissenters" appear to be more numerous than Martin supposes. (We have certainly seen that Ernest Thompson Seton's religious views were anything but "mainstream," and in fact he disavowed Christianity at an early age [Seton 1940, 137].) Calvin Martin believes that it is essentially impossible for "Western man" to accept or adopt Native American attitudes toward the environment, "at least as long as he remains a Christian" (Martin 1978, 188 n.e). This may be true—but the evidence suggests that "Western man" may not remain a Christian (or at least, a strict monotheist) as consistently as Martin supposes. Environmentalists in particular appear to incline toward other forms of faith.

It may yet be objected that even for environmentalists who have rejected Christian monotheism, Native American traditions must remain "alien" and unacceptable. Martin uses an example from his Eastern Algonkian researches to illustrate this point. Algonkian hunters traditionally killed only to fulfill their subsistence needs. According to Martin, this was primarily because they feared that the animals would take revenge for unnecessary killing by using their supernatural powers to inflict disease upon human beings (Martin 1978, 18, 73–74). Surely, Martin argues, no modern Westerner could accept such an idea.

This argument is not without force. Even Ernest Thompson Seton would scarcely have accepted the Algonkian belief that animals could use magic to cause human illness. Is it not possible, however, for someone to "accept" a particular religious worldview without necessarily accepting every specific belief associated with it? Martin, for instance, presumes that virtually all Westerners "accept" the worldview of Christianity. But he cannot possibly imagine that every Westerner accepts such traditional Christian teachings as that disease is caused by malignant spirits that can be exorcised through prayer. Obviously, few Westerners today would accept this idea, though it is undeniably basic to the outlook of the New Testament. Yet Martin argues, correctly I think, that most Westerners nonetheless retain a worldview that is essentially "Christian" in content.

In a similar manner, I would argue, some environmentalists have consciously and deliberately adopted a worldview that is essentially "Native American" in content, which is not to say that any of them have adopted in toto the beliefs of any particular native people. While none might accept the Algonkian notion that animals can magically cause disease, some do appreciate and accept the worldview that makes such a notion possible—a worldview in which animals and humanity are seen as interdependent parts of a single ecological system.

In many ways the Algonkian theory of disease is a perfect symbol for what Barry Commoner has called the first law of ecology: "Everything is connected to everything else." Because this is so, environmentalists argue, humanity cannot endlessly exploit nature without causing repercussions that we may ultimately regret. The ecosystem is an integrated web of cause and effect from which we cannot extricate ourselves. Anything we do to any aspect of nature affects the web and so affects us. Thus the animals *will* ultimately have revenge for their overexploitation by humanity—if not through disease, then through some other consequence of the unbalancing of the ecosystem. Environmentalists may not accept Algonkian ideas about disease as literal fact, but the fundamental truth about the nature of reality that those ideas symbolize is central to the philosophy of the environmental movement.

American environmentalists have admired Native American traditions since the inception of the movement, and it is no surprise that contemporary environmentalists find themselves attracted to and inspired by native values. Those values have to some extent made the American environmental movement what it is. Many of the individuals who helped shape the present philosophy of the movement were themselves shaped by their knowledge of and respect for Native American traditions. The topic surely needs exploration commensurate with its scope and importance.

NOTES

1. Cf. Julia M. Seton's introduction to the 1937 edition of *The Gospel of the Redman* (reprinted in the 1966 edition).

2. Seton published a new edition of "Birch Bark Roll" annually for many years.

3. According to Ms. Barber, there is also a remnant of the league itself still active in the Los Angeles area.

REFERENCES

Brochure for "a school of the woods" (later called Camp Oahe). 1915. Eastman Collection, Jones Library, Amherst, Mass.

Brochure for Camp Ohiyesa. 1917. Eastman Collection, Jones Library, Amherst, Mass.

Callicott, J. Baird. 1982. Traditional American Indian and Western European attitudes toward nature: An overview. *Environmental Ethics* 4(4): 313–14.

Doughty, Howard. 1962. *Francis Parkman*. New York: Macmillan.

Eastman, Charles Alexander [Ohiyesa]. 1980. *The soul of the Indian: An interpretation*. Lincoln: University of Nebraska Press. Originally published 1911.

Festinger, L. 1957. *A theory of cognitive dissonance*. Evanston, Ill.: Row, Peterson.

Fox, Stephen. 1981. *John Muir and his legacy: The American conservation movement*. Boston: Little, Brown.

Krech, Shepard, III, ed. 1981. *Indians, animals and the fur trade: A critique of "Keepers of the Game."* Athens: University of Georgia Press.

Levy, Harold P. 1944. *Building a popular movement: A case study of the public relations of the Boy Scouts of America*. New York: Russell Sage Foundation.

Maloney, H. Newton. 1976. New methods in the psychology of religion. *Journal of Psychology and Theology* 4(2): 141–51.

Martin, Calvin. 1978. *Keepers of the Game: Indian-animal relationships and the fur trade*. Berkeley: University of California Press.

———. 1981. The American Indian as miscast ecologist. In *Ecological consciousness*, ed. J. Donald Hughes and Robert C. Schultz. Washington, D.C.: University Press of America.

Miller, David R. 1978. Charles Alexander Eastman, the "Winner": From deep woods to civilization. In *American Indian intellectuals*, ed. Margot Liberty. St. Paul, Minn.: West.

Nash, Roderick, ed. 1976. *The American environment: Readings in the history of conservation*. 2d ed. Reading, Mass.: Addison-Wesley.

Parkman, Francis. 1964. *The Oregon Trail*. New York: Dodd, Mead. Originally published 1849.

Regan, Tom. 1982. *All that dwell therein: Animal rights and environmental ethics*. Berkeley: University of California Press.

Sayre, Robert F. 1977. Thoreau and the American Indians. Princeton, N.J.: Princeton University Press.

Seton, Ernest Thompson. 1898. *Wild animals I have known*. New York: Random House.

———. 1923. *The book of woodcraft and Indian lore*. 11th Birch Bark Roll. Garden City, N.Y.: Doubleday.

———. 1940. *Trail of an artist-naturalist: The autobiography of Ernest Thompson Seton*. New York: Charles Scribner's Sons.

———. 1954. The Buffalo wind. In *Ernest Thompson Seton's America: Selections from the writings of the artist-naturalist*, ed. Farida Wiley. New York: Devin-Adair.

Seton, Ernest Thompson, and Julie M. Seton. 1966. *The gospel of the redman: A way of life*. Santa Fe, N.Mex.: Seton Village. Originally published 1938.

Seton, Julia M. 1967. *By a thousand fires: Nature notes and extracts from the life and unpublished journal of Ernest Thompson Seton*. Garden City, N.Y.: Doubleday.

Steinhart, Peter. 1984. Ecological saints. *Audubon* 86:4.

Strong, Douglas H. 1971. *The conservationists*. Menlo Park, Calif.: Addison-Wesley.

Udall, Stewart L. 1963. *The quiet crisis*. New York: Holt, Rinehart and Winston.

Wild, Peter. 1979. *Conservationists of western America*. Missoula, Mont.: Mountain Press.

Conclusion:
Two Parables and a Reprise

Paul A. Olson

Two "parables" of sustainability and destruction in semiarid regions may be offered here. The first concerns the Ngisonyoka Turkana from the area west and south of Lake Turkana in Kenya, near the Maasai areas that Soloman Bekure and Ishmael Ole Pasha discuss. In a recent article, J. Terrence Mc-Cabe and James E. Ellis pointed out that in the 1970s the main body of Turkana pastoralists were encouraged to adopt irrigation agriculture and fish culture as an alternative to their traditional pastoral nomadic way of life. When drought and disease prevented the fields and lakes from producing their promised abundance, the people who had given up the old way were forced into famine camps. On the other hand, one group of Turkana pastoralists, the Ngisonyoka, remained a self-supporting, viable group in the same semiarid region by moving their herds from the wet-season grassy plains of the Rift Valley to nearby lava hills and mountains that furnished dry-season perennial grass forage. The group's mixed array of camels, oxen, goats, and sheep ate over sixty kinds of plants in the surrounding ecosystem, and when the dry season came the herds were simply moved to new forage. As the animals moved, the human diet changed from the animal meat abundant in the rainy season to the dry-season staples of camel's milk, livestock blood, wild fruits, goat meat, and grain purchased from the outside. As the forage became scarcer, the herding groups became smaller, but their networks of interdependence became more complex to allow for "sharing the effects of drought stress among families. . . . The Ngisonyoka have worked out a strategy: they depend most on the most reliable resources in their environment—woody plants and camel milk; they exploit the most productive but ephemeral resources—grasses and cattle milk—when possible; when times are hard, they make use of their precious livestock—drinking its blood, slaughtering it for meat, or trading it for grain. These are never arbitrary decisions. For example, goats are most often slaughtered and traded because goat herds recover most rapidly" (McCabe and Ellis 1987, 39). The Ngisonyoka do not irrigate because irrigation costs more than it pays in their system; they do not raise cattle for the international market because they know they need their cattle for times of emergency. They have endured, as has their surrounding world.

The second parable, related by Farley Mowat in his *People of the Deer*, tells of the recent destruction of the Caribou Inuit, who lived inland from Hudson Bay, through the Canadian decimation of the herds they hunted—a story in its way as sordid as the destruction of the buffalo and of the autonomy of the Plains tribes in the nineteenth century. Mowat tells of one of the members of the group who, on his vision quest, dreamed of a monster white man that struggled with him and conquered. The vision seeker then set off across the land and reached a northerly interior white outpost where he acquired many of the trinkets of European civilization, including guns. When he returned, his fellow tribesmen took his trinkets to examine them. One of his brothers took a gun to examine it. Afraid to lose a prized possession, the vision seeker slashed his brother's shoulder with an ax, recovered the rifle, and left the tribe to live as a solitary and pillage his fellow tribesmen of their pots, women, or whatever. He lived alone surrounded by rusting apparatus that he could not use and was described by his own people as the Inuit who lived as white men do (Mowat 1952, 201–34).

Most of the population of the semiarid world, whether aboriginal or European, can no longer follow the path of the Ngisonyoka. Though the "monster white man" and his civilization may be feared or despised for their waste, their system has taken over.

The problems described in the book exist at several levels, extending from what structuralists describe as infrastructure to superstructure: compatability of technology and resources exploited with sustainability and conservation; resource access; legal protection for sustainable lifeways; values and ethical-religious structures; and structure of the disciplines.

Compatability of resources exploited with sustainability and conservation: The general direction of the studies in this book suggests that plants and animals that have evolved in an area, or have been bred there for long periods, have a better chance of flourishing than others, that resource management practices that retain cover or require minimal tillage or tillage in limited areas are most likely to sustain themselves in semiarid regions. Such regimens have often been developed in semiarid regions by indigenes or European-based people separate from the system of the world market.

Resource access: The predominant tendency of the European industrial system has been to move access to resources out of the hands of subsistence-based local peoples and into the centers of the world system and to require that products created "fit" the world market whether or not they flourish in the local region. This usually requires high imports of energy, political muscle, and industrial technology on the part of the colonizing power, which in turn extracts products from the region at a high cost to its soil and its water and mineral resources. As pressure to "produce more" develops in semiarid regions, the lands set aside for indigenes as reserves may be alienated from their control through allotment, the imposition of dams and mines required for the "national" interest, and the formation of "native corporations," or through efforts to assimilate indigenes without essentially changing their second-class citizenship in the nation-state.

Legal protection for sustainable lifeways: Efforts to preserve the indigenous group and protect it from gross exploitation and total loss of resources are usually conducted by the European-based nation-state on its terms, under rubrics already contained in its laws. No serious international protection exists for indigenous groups, and there are few effectual models of the creation of workable intercultural governance systems. This problem manifests itself whether the indigenous group resides in a First World nation such as the United States, a Second World nation such as the USSR, or a Third World nation such as Kenya. If tribal people's practices are, in any serious way, to inform or model development, they will have to be protected.

Values and ethical/religious structures: The role of ethical/religious structures in conservation is not clear, partly because few detailed analyses have been done of the use of stories, myths, and proverbs in the context of children's and adult's actual behavior toward the environment. Certainly one can demonstrate, as Callicott has, that some ethical/religious systems are conservationist in content and some are not. John Bennett has argued that

> the environment has, in most human history up to now, been viewed as a reservoir of raw materials—resources—for serving human ends, and only in tribal societies, with their limited population base, have such views coincided with conservationist strategies, or at last, minimal pressure on Nature. However, there is little or no evidence in the ethnological record that any group of humans ever consciously and conscientiously sought to husband resources. Most such instances refer instead to values or rituals which rationalize or validate existing minimal usage due to lack of population pressure. Modest needs beget modest ideologies; it is a kind of "low-level equilibrium trap" and humans have shown little tolerance of it in the long haul. (Bennett 1986, 348–49)

However, the semiarid environment has not tolerated high-level exploitation, and we have no models of what high-level semiarid equilibrium would be like.

Structure of the disciplines: This book was conceived as an interdisciplinary examination because its editor wished to look at a series of cases of what has happened to the environment and to people in cases where indigenous peoples living in semiarid regions have been replaced by industrial "man." Both its composition and its reviewing have convinced me that the disciplines as now organized are not adequately prepared to study the questions posed here. We are just beginning to look at the problems of semiarid regions as part of the world environmental system and in relationship to individual nation-state and tribal ideologies and technologies. Anthropology tends to look at micro issues, neoclassical economics at the world system apart from human institutions and actions, history at the actions of writing-centered societies that can furnish documentation, religion and philosophy at human ideology separated from day-to-day actions and teachings, and literature and philosophy at texts divorced from actions. Such an organization of study is not likely to lead to real comprehension or effective policy. In the absence of comprehension and conscience, brute force is apt to triumph as we endeavor to put together a world human order that respects the other orders of nature and has some potential for choosing a sustainable relationship with it.

REFERENCES

Bennett, John W. 1986. Summary and critique: Interdisciplinary research on people-resource relations. In *Natural Resources and people: Conceptual issues in interdisciplinary Research*, 343-72. Boulder, Colo.: Westview Press.

McCabe, J. Terrence, and James E. Ellis. 1987. Beating the odds in arid Africa. *Natural History* 96(1): 32–41.

Mowat, Farley. 1952. *People of the deer*. Boston: Little, Brown.

The Contributors

Gary C. Anders, a Cherokee, is professor of economics at the Arizona State University West Campus. He has worked on economic results of government policies toward Alaska Natives and has published articles on development policy and Native Americans in various scholarly journals and books.

Russel L. Barsh is a lawyer and former faculty member at the University of Washington. His publications in ecology and law include *Karluk River Study* (1985) and *Seals and Sealing in Canada* (1986) as well as articles in various scholarly journals. He recently worked on the adoption by the United Nations International Labour Organisation of a convention on indigenous and tribal peoples, which recognizes the right of these peoples to manage their own natural resources.

Solomon Bekure is employed with the World Bank and has worked with the International Livestock Centre for Africa in Kenya, the Ministry of Agriculture in Ethiopia, the Agricultural and Industrial Development Bank in Ethiopia, and the Faculty of Agriculture at Haile Selassie University. He has written numerous articles on Ethiopian and Kenyan agriculture and pastoralism.

John W. Bennett is a professor of anthropology and is also affiliated with the East Asian Center and the Technology and Human Affairs program at

Washington University, St. Louis. He specializes in ecological and agrarian development problems. His recent publications may be surveyed in the bibliography of his essay.

J. Baird Callicott is a professor of environmental studies at the University of California, Santa Barbara. His publications include *Companion to a Sand County Almanac: Interpretative and Critical Essays* (1988) and *Environmental Philosophy: The Nature of Nature in Asian Traditions of Thought* (1989) and other books. He has written numerous essays on subjects joining philosophy and the environmental sciences and is on the editorial board of several scholarly journals.

Annette Hamilton is a professor of anthropology and comparative sociology at Macquarie University, Sydney. She has worked extensively on projects concerned with Australian aboriginal land claims and social and economic development. She is also doing research in Thailand on the relationship between forms of communication and socioeconomic transformation. She is the author of *Nature and Nurture: Child Rearing in North-Central Arnhem Land* (1980) and numerous papers.

Peter Iverson is a professor of history at Arizona State University, Tempe. His publications include *Carlos Montezuma and the Changing World of American Indians* (1982) and *The Plains Indians of the Twentieth Century* (1985) as well as other books and articles in scholarly journals.

Anatoly Khazanov is a professor in the Department of Sociology and Social Anthropology at the Hebrew University, Jerusalem. He emigrated from the Soviet Union in 1985 and has done fieldwork in the Ukraine, Central Asia, Daghestan, Latvia, Kalmykia, and the Caucasus. His publications include *The Primitive Periphery of Class Societies* (1978) and *Nomads of the Outside World* (1984), as well as other books and numerous articles on issues in historical anthropology.

C. Patrick Morris is a professor at the Center for Native American Studies at Montana State University, Bozeman. He has contributed chapters for several books and is active in a variety of programs designed to protect the rights of indigenous people.

Paul Olson is Foundation Professor of English at the University of Nebraska. He has taught Native American literature for more than twenty years and has advised the Lincoln Indian Center, the Nebraska State Indian Commission, and the Northern Cheyenne tribe in various practical economic development activities. His writings include *The Book of the Omaha* (1979) and *A Few Great Santee Stories* (1979).

Ishmael Ole Pasha, a Maasai, has served as a civil servant with the government of Kenya and as a research officer with the International Livestock Center for Africa, concentrating especially on research activities in Maasailand.

O. Douglas Schwarz has taught religion at Vassar College and is at present a private scholar. He has published articles on environmental ethics and Native American theology in various scholarly journals.

Robson Silitshena is senior lecturer in the environmental sciences department at the University of Botswana. He is the author of *Intra-mural Migration and Settlement Changes in Botswana* (1983) and has contributed chapters to books, articles in various scholarly journals, and a number of reports for the government of Botswana. He is conducting land-use and development planning studies for the Kalahari Conservation Society.

Index